TRIUMPH TR2-TR3 1952-1960

Compiled by
R.M. Clarke

ISBN 1 870642 589

Distributed by
Brooklands Book Distribution Ltd.
'Holmerise', Seven Hills Road,
Cobham, Surrey, England

Printed in Hong Kong

BROOKLANDS BOOKS

BROOKLANDS BOOKS SERIES
AC Ace & Aceca 1953-1983
AC Cobra 1962-1969
Alfa Romeo Alfasud 1972-1984
Alfa Romeo Alfetta Coupes GT.GTV.GTV6 1974-1987
Alfa Romeo Guilias Berlinettas
Alfa Romeo Giulia Berlinas 1962-1976
Alfa Romeo Giulia Coupés 1963-1976
Alfa Romeo Spider 1966-1987
Allard Gold Portfolio 1937-1958
Aston Martin Gold Portfolio 1972-1985
Austin Seven 1922-1982
Austin A30 & A35 1951-1962
Austin Healey 100 1952-1959
Austin Healey 3000 1959-1967
Austin Healey 100 & 3000 Collection No. 1
Austin Healey 'Frogeye' Sprite Collection No. 1
Austin Healey Sprite 1958-1971
Avanti 1962-1983
BMW Six Cylinder Coupés 1969-1975
BMW 1600 Collection No. 1
BMW 2002 1968-1976
Bristol Cars Gold Portfolio 1946-1985
Buick Automobiles 1947-1960
Buick Riviera 1963-1978
Cadillac Automobiles 1949-1959
Cadillac Automobiles 1960-1969
Cadillac Eldorado 1967-1978
Camaro 1966-1970
Chevrolet Camaro & Z-28 1973-1981
High Performance Camaros 1982-1988
Chevrolet Camaro Collection No. 1
Chevrolet 1955-1957
Chevrolet Impala & SS 1958-1971
Chevelle & SS 1964-1972
Chevy II Nova & SS 1962-1973
Chrysler 300 1955-1970
Citroen Traction Avant 1934-1957
Citroen DS & ID 1955-1875
Citroen 2CV 1948-1988
Cobras & Replicas 1962-1983
Cortina 1600E & GT 1967-1970
Corvair 1959-1968
Daimler Dart & V-8 250 1959-1969
Datsun 240z 1970-1973
Datsun 280Z & ZX 1975-1983
De Tomaso Collection No. 1
Dodge Charger 1966-1974
Excalibur Collection No. 1
Ferrari Cars 1946-1956
Ferrari Cars 1962-1966
Ferrari Cars 1969-1973
Ferrari Dino 1965-1974
Ferrari Dino 308 1974-1979
Ferrari 308 & Mondial 1980-1984
Ferrari Collection No. 1
Fiat-Bertone X1/9 1973-1988
Fiat Pininfarina 124+2000 Spider 1968-1985
Ford Falcon 1960-1970
Ford Mustang 1964-1967
Ford Mustang 1967-1973
High Performance Mustangs 1982-1988
Ford RS Escort 1968-1980
Honda CRX 1983-1987
High Performance Escorts MkI 1968-1974
High Performance Escorts MkII 1975-1980
Hudson & Railton Cars 1936-1940
Jaguar XK120 XK140 XK150 Gold Portfolio 1948-1960
Jaguar Cars 1957-1961
Jaguar Cars 1961-1964
Jaguar MK2 1959-1969
Jaguar E-Type 1961-1966
Jaguar E-Type 1966-1971
Jaguar E-Type V12 1971-1975
Jaguar XKE Collection No. 1
Jaguar XJ6 1968-1972
Jaguar XJ6 Series II 1973-1979
Jaguar XJ6 & XJ12 Series III 1979-1985
Jaguar XJ12 1972-1980
Jaguar XJS Gold Portfolio 1975-1988
Jensen Cars 1946-1967
Jensen Cars 1967-1979
Jensen Interceptor Gold Portfolio 1966-1986
Lamborghini Cars 1964-1970
Lamborghini Cars 1970-1975
Lamborghini Countach Collection No. 1
Lamborghini Countach & Urraco 1974-1980
Lamborghini Countach & Jalpa 1980-1985
Lancia Stratos 1972-1985
Land Rover 1948-1973
Land Rover Series II & IIa 1958-1971
Land Rover Series III 1971-1985
Land Rover 90 & 110 1983-1989
Lotus Cortina 1963-1970
Lotus Elan Gold Portfolio 1962-1974
Lotus Elan Collection No. 2
Lotus Elite 1957-1964
Lotus Elite & Eclat 1974-1981
Lotus Turbo Esprit 1980-1986
Lotus Europa 1966-1975
Lotus Europa Collection No. 1
Lotus Seven 1957-1980
Lotus Seven Collection No. 1
Maserati 1965-1970
Maserati 1970-1975
Marcos Cars 1960-1988
Mazda RX-7 Collection No. 1
Mercedes 190 & 300SL 1954-1963
Mercedes 230/250/280SL 1963-1971
Mercedes 350/450SL & SLC 1971-1980
Mercedes Benz Cars 1949-1954
Mercedes Benz Cars 1954-1957
Mercedes Benz Cars 1957-1961
Mercedes Benz Competition Cars 1950-1957

Metropolitan 1954-1962
MG Cars 1929-1934
MG TC 1945-1949
MG TD 1949-1953
MG TF 1953-1955
MG Cars 1957-1959
MG Cars 1959-1962
MG Midget 1961-1980
MGA Collection No. 1
MGA Roadsters 1955-1962
MGB Roadsters 1962-1980
MGB GT 1965-1980
Mini Cooper 1961-1971
Morgan Cars 1960-1970
Morgan Cars 1969-1979
Morris Minor Collection No. 1
Old's Cutlass & 4-4-2 1964-1972
Oldsmobile Toronado 1966-1978
Opel GT 1968-1973
Packard Gold Portfolio 1946-1958
Pantera 1970-1973
Pantera & Mangusta 1969-1974
Plymouth Barracuda 1964-1974
Pontiac Fiero 1984-1988
Pontiac Firebird 1967-1973
Pontiac Firebird and Trans-Am 1973-1981
High Performance Firebirds 1982-1988
Pontiac Tempest & GTO 1961-1965
Porsche Cars 1960-1964
Porsche Cars 1964-1968
Porsche Cars 1968-1972
Porsche Cars in the Sixties
Porsche Cars 1972-1975
Porsche 356 1952-1965
Porsche 911 Collection No. 1
Porsche 911 Collection No. 2
Porsche 911 1965-1969
Porsche 911 1970-1972
Porsche 911 1973-1977
Porsche 911 Carrera 1973-1977
Porsche 911 SC 1978-1983
Porsche 911 Turbo 1975-1984
Porsche 914 Gold Portfolio 1969-1988
Porsche 914 Collection No. 1
Porsche 924 1975-1981
Porsche 928 Collection No. 1
Porsche 944 1981-1985
Porsche Turbo Collection No. 1
Reliant Scimitar 1964-1986
Riley 1½ & 2½ Litre Gold Portfolio 1945-1955
Rolls Royce Silver Cloud 1955-1965
Rolls Royce Silver Shadow 1965-1980
Range Rover Gold Portfolio 1970-1988
Rover 3 & 3.5 Litre 1958-1973
Rover P4 1949-1959
Rover P4 1955-1964
Rover 2000 + 2200 1963-1977
Rover 3500 1968-1977
Rover 3500 & Vitesse 1976-1986
Saab Sonett Collection No. 1
Saab Turbo 1976-1983
Studebaker Hawks & Larks 1956-1963
Sunbeam Tiger And Alpine Gold Portfolio 1959-1967
Thunderbird 1955-1957
Thunderbird 1958-1963
Thunderbird 1964-1976
Toyota MR2 1984-1988
Triumph 2000-2.5-2500 1963-1977
Triumph Spitfire 1962-1980
Triumph Spitfire Collection No. 1
Triumph Stag 1970-1980
Triumph Stag Collection No. 1
Triumph TR2 & TR3 1952-1960
Triumph TR4.TR5.TR250 1961-1968
Triumph TR6 1969-1976
Triumph TR6 Collection No. 1
Triumph TR7 & TR8 1975-1982
Triumph GT6 1967-1974
Triumph Vitesse & Herald 1959-1971
TVR Gold Portfolio 1959-1988
Volkswagen Cars 1936-1956
VW Beetle 1956-1977
VW Beetle Collection No. 1
VW Golf GTi 1976-1986
VW Karmann Ghia 1955-1982
VW Scirocco 1974-1981
VW Bus-Camper-Van 1954-1967
VW Bus-Camper-Van 1968-1979
Volvo 1800 1960-1973
Volvo 120 Series 1956-1970

BROOKLANDS MUSCLE CARS SERIES
American Motors Muscle Cars 1966-1970
Buick Muscle Cars 1965-1970
Camaro Muscle Cars 1966-1972
Capri Muscle Cars 1969-1983
Chevrolet Muscle Cars 1966-1972
Dodge Muscle Cars 1967-1970
Mercury Muscle Cars 1966-1971
Mini Muscle Cars 1961-1979
Mopar Muscle Cars 1964-1967
Mopar Muscle Cars 1968-1971
Mustang Muscle Cars 1967-1971
Shelby Mustang Muscle Cars 1965-1970
Oldsmobile Muscle Cars 1964-1970
Plymouth Muscle Cars 1966-1971
Pontiac Muscle Cars 1966-1971
Muscle Cars Compared 1966-1971
Muscle Cars Compared Book 2 1965-1971

BROOKLANDS ROAD & TRACK SERIES
Road & Track on Alfa Romeo 1949-1963
Road & Track on Alfa Romeo 1964-1970
Road & Track on Alfa Romeo 1971-1976

Road & Track on Alfa Romeo 1977-1984
Road & Track on Aston Martin 1962-1984
Road & Track on Auburn Cord & Duesenberg 1952-1984
Road & Track on Audi 1952-1980
Road & Track on Audi 1980-1986
Road & Track on Austin Healey 1953-1970
Road & Track on BMW Cars 1966-1974
Road & Track on BMW Cars 1975-1978
Road & Track on BMW Cars 1979-1983
Road & Track on Cobra, Shelby &
 Ford GT40 1962-1983
Road & Track on Corvette 1953-1967
Road & Track on Corvette 1968-1982
Road & Track on Corvette 1982-1986
Road & Track on Datsun Z 1970-1983
Road & Track on Ferrari 1950-1968
Road & Track on Ferrari 1968-1974
Road & Track on Ferrari 1975-1981
Road & Track on Ferrari 1981-1984
Road & Track on Fiat Sports Cars 1968-1987
Road & Track on Jaguar 1950-1960
Road & Track on Jaguar 1961-1968
Road & Track on Jaguar 1968-1974
Road & Track on Jaguar 1974-1982
Road & Track on Jaguar 1983-1989
Road & Track on Lamborghini 1964-1985
Road & Track on Lotus 1972-1981
Road & Track on Maserati 1952-1974
Road & Track on Maserati 1975-1983
Road & Track on Mazda RX7 1978-1986
Road & Track on Mercedes 1952-1962
Road & Track on Mercedes 1963-1970
Road & Track on Mercedes 1971-1979
Road & Track on Mercedes 1980-1987
Road & Track on MG Sports Cars 1949-1961
Road & Track on MG Sports Cars 1962-1980
Road & Track on Mustang 1964-1977
Road & Track on Peugeot 1955-1986
Road & Track on Pontiac 1960-1983
Road & Track on Porsche 1951-1967
Road & Track on Porsche 1968-1971
Road & Track on Porsche 1972-1975
Road & Track on Porsche 1975-1978
Road & Track on Porsche 1979-1982
Road & Track on Porsche 1982-1985
Road & Track on Rolls Royce & Bentley 1950-1965
Road & Track on Rolls Royce & Bentley 1966-1984
Road & Track on Saab 1955-1985
Road & Track on Toyota Sports & G T Cars 1966-1986
Road & Track on Triumph Sports Cars 1953-1967
Road & Track on Triumph Sports Cars 1967-1974
Road & Track on Triumph Sports Cars 1974-1982
Road & Track on Volkswagen 1951-1968
Road & Track on Volkswagen 1968-1978
Road & Track on Volkswagen 1978-1985
Road & Track on Volvo 1957-1974
Road & Track on Volvo 1975-1985
Road & Track Henry Manney At Large & Abroad

BROOKLANDS CAR AND DRIVER SERIES
Car and Driver on BMW 1955-1977
Car and Driver on BMW 1977-1985
Car and Driver on Cobra, Shelby & Ford GT40
 1963-1984
Car and Driver on Datsun Z 1600 & 2000
 1966-1984
Car and Driver on Corvette 1956-1967
Car and Driver on Corvette 1968-1977
Car and Driver on Corvette 1978-1982
Car and Driver on Corvette 1983-1988
Car and Driver on Ferrari 1955-1962
Car and Driver on Ferrari 1963-1975
Car and Driver on Ferrari 1976-1983
Car and Driver on Mopar 1956-1967
Car and Driver on Mopar 1968-1975
Car and Driver on Mustang 1964-1972
Car and Driver on Pontiac 1961-1975
Car and Driver on Porsche 1955-1962
Car and Driver on Porsche 1963-1970
Car and Driver on Porsche 1970-1976
Car and Driver on Porsche 1977-1981
Car and Driver on Porsche 1982-1986
Car and Driver on Saab 1956-1985
Car and Driver on Volvo 1955-1986

BROOKLANDS MOTOR & THOROUGHBRED & CLASSIC CAR SERIES
Motor & T & CC on Ferrari 1966-1976
Motor & T & CC on Ferrari 1976-1984
Motor & T & CC on Lotus 1979-1983
Motor & T & CC on Morris Minor 1948-1983

BROOKLANDS PRACTICAL CLASSICS SERIES
Practical Classics on Austin A 40 Restoration
Practical Classics on Land Rover Restoration
Practical Classics on Metalworking in Restoration
Practical Classics on Midget/Sprite Restoration
Practical Classics on Mini Cooper Restoration
Practical Classics on MGB Restoration
Practical Classics on Morris Minor Restoration
Practical Classics on Triumph Herald/Vitesse
Practical Classics on Triumph Spitfire Restoration
Practical Classics on VW Beetle Restoration
Practical Classics on 1930S Car Restoration

BROOKLANDS MILITARY VEHICLES SERIES
Allied Military Vehicles Collection No. 1
Allied Military Vehicles Collection No. 2
Dodge Military Vehicles Collection No. 1
Military Jeeps 1941-1945
Off Road Jeeps 1944-1971
V W Kubelwagen 1940-1975

CONTENTS

5	A New 2-Litre Triumph Sports Car	Motor	Oct. 22	1952
8	A Triumph Sports Model	Autocar	Oct. 22	1952
11	The Birth of a Sports Car	Motor	May 27	1953
14	1954 Cars – Triumph	Motor	Oct. 21	1953
15	Triumph Sports Achieves 124 MPH	Motor	May 27	1953
16	The Price of Speed	Autocar	June 5	1953
18	The 120 MPH Triumph	Light Car	July	1953
20	The Triumph TR2 Road Test	Autosport	Mar. 5	1954
24	A Taste of Triumph	Autocar	Sept. 17	1954
27	1955 Cars – Triumph	Motor	Oct. 22	1954
28	The Triumph Sports 2-Seater Road Test	Motor	April 7	1954
32	A Triumph of Development	Autocar	April 8	1954
40	From One Generation to Another	Motor	June 22	1955
44	A TR2 – Plus	Autosport	Sept. 16	1955
45	Trying the Triumph	Sports Cars Illustrated	Nov.	1955
48	The Triumph TR3 Hard-top Coupe Road Test	Motor	April 4	1956
52	Test of the TR3 Road Test	Sports Cars Illustrated	Mar.	1956
56	Disc Brakes for the TR3	Autocar	Sept. 28	1956
58	The Gallant Mountaineer	Autocar	July 27	1956
60	Triumph TR3 Road Test	Top Gear		1957
63	A Tale of Two-Seaters	Motor	Jan. 16	1957
68	The Triumph TR3 Hardtop Road Test	Motor	July 3	1957
72	Triumph TR3 Road Test	Motor Trend	Sept.	1957
74	Triumph TR3 Road Test	Autosport		1957
77	An Improved Triumph TR3	Motor	Jan. 1	1958
78	Triumph TR3 Road Test	Sports Cars Illustrated	Mar.	1958
80	Model in Many Ways	Autocar	Aug. 22	1958
82	Triumph TR3 Hard Top	Sports Cars Illustrated	Feb.	1958
85	1955 Triumph TR2 Hard Top	Autocar	Mar. 28	1958
86	Disc Braked TR3 Road Test	Sports Cars Illustrated	Mar.	1958
91	Buying a TR2, TR3 or TR3A	Practical Classics	Oct.	1982
96	TR Triumphant	Motor	April 8	1959
100	Joyful and Triumphant Profile	Classic and Sportscar	Aug.	1986

Road & Track articles and road tests can be found in our publication 'Road & Track on Triumph Sportscars' 1953-1967

ACKNOWLEDGEMENTS

Production of the Triumph Roadster came to an end in 1949 and the story goes that sometime later Sir John Black, who was managing director of the Standard Motor Company, of which Triumph was a subsidary, wanted a sports car in the range. It is said that he was jealous of Nuffield who were shipping the MG TDs to the United States by the boat-load and was also a little peeved with his neighbour William Lyons who pulled the fabulously successful XK120 out of his hat in 1948.

He had already tried one short cut to achieve his aim by making a proposal to his old friend in Malvern. H.S.F. Morgan who was producing the Standard engined Morgan Four 4 which enjoyed then and still does an enviable reputation. The marriage however was not to be.

Luckily for Sir John, Standard and Triumph had between them the ideal ingredients for a British sports car, there were some left-over chassis from the Flying Standard Nine, a selection of parts from the Mayfower mini-limosine and an unburstable converted tractor engine that had been pushing around numerous Vanguards since 1948, plus the miracle ingredient a competent staff.

Out of this emerged the TR1 in 1952 which was immediately pruned and went on to flower in May the following year by flying down the Ostend-Ghent motorway at 124 mph. The TR2 went into production three months later. By the end of 1954 it had won the R.A.C. Rally outright and completed the 24-hour race at Le Mans at an average speed of 74.71 mph. It grew into the TR3 in October 1955 and went on to sire a proud line of sportscars.

Brooklands Books are a reference series for owners, historians, restorers and car lovers in general and exist because for over 30 years the publishers of the worlds leading motoring journals have generously allowed us to reissue their copyright road tests and other factual motoring stories.

I am sure that Triumph devotees will wish to join us in thanking the management of Autocar, Autosport, Classic and Sportscar, Sports Car Illustrated US, Sports Car Illustrated UK, Light Car, Motor, Motor Trend, Practical Classics and Top gear for their ongoing support.

R.M. Clarke

October 22, 1952

THE MOTOR

1953 CARS

Combination of Light Two-seater Structure with 75 b.h.p. Engine Gives Promise of Lively Performance, and at Only £555 Basic Price

A New 2-litre Triumph Sports Car

AN increase in popular enthusiasm for motor sport in England was a natural post-war corollary of the wide dispersion of mechanical knowledge throughout members of the armed forces during the wartime years, and it was reasonable to predict a correspondingly enlarged market for cars of sporting type. But probably no one anticipated that this expansion would be most marked in overseas countries, and especially in the U.S.A. It has, however, turned out so to be.

Since 1946 there have not only been record numbers of spectators at motor races and record numbers of entrants in rallies, hill climbs and competitions of all kinds, but also a dramatic demand for sports cars in the United States, initially centred in California, but now supplemented from the East, and even from Detroit itself.

In pre-war years the Triumph Motor Co. gained many successes with the sporting cars in their range of models, especially in the Alpine Trial and Monte-Carlo Rally. In the light of post-war events it was therefore not unexpected that the Triumph range should again include a sports car, and it is interesting to see that this new design is designed to give a very high performance in the hands of the ordinary user, to be of the simplest possible construction, and to sell at a moderate price.

Reference to the specification shows that the theme of the design is the installation of a comparatively large engine in the smallest and lightest practicable motorcar. The wheelbase of 7 ft. 4 ins. is, in particular, exceptionally short, and a 3 ft. 9 ins. track is also a little less than is normal to this type of car. The overall dimensions reflect a certain amount of overhang, but the very compact grouping of the mechanical structure has resulted in an exceptionally low estimated weight of 15¼ cwt., and even should this be exceeded slightly in the production models, an all-up weight, with driver and passenger, of one ton appears well within the realms of possibility. Even on this perhaps pessimistic estimate the car will dispose of 75 h.p./laden ton, displace 3,100 litres per ton-mile on top gear and have over 120 sq. ins. per ton of friction lining area. Figures of this order will provide acceleration and hill-climbing of a very high order despite the use of a top gear which permits a cruising speed of 80 m.p.h. at 2,500 ft./min. and an estimated maximum speed of 90 m.p.h.

These figures derive from the use of a short-stroke, two-litre engine (83 by 96 mm.) which is a developed version of the four-cylinder power unit used currently in the Triumph Renown and previously also employed in the Triumph Roadster.

In the 2-litre Class

As is well known, a distinctive feature of this engine is quickly detachable wet cylinder liners and by adding 1 mm. to the wall thickness of these sleeves the swept volume of the engine has been decreased from 2,088 c.c. to 1,991 c.c. (with a corresponding reduction in piston area) so as to bring the car within the 2-litre class for competition events.

The manifolding of the engine has been modified in that a single downdraught fixed-

TRIUMPH SPORTS

Engine Dimensions		Transmission—Contd.	
Cylinders	4	Prop. shaft	Hardy Spicer
Bore	83 mm.	Final drive	Hypoid bevel
Stroke	96 mm.	**Chassis Details**	
Cubic capacity	1,991 c.c.	Brakes	Lockheed
Piston area	33.5 sq. ins.	Brake drum diameter	9 ins.
Valves	pushrod o.h.v.	Friction lining area	121 sq. ins.
Compression ratio	7 to 1	Suspension: Front	Open coil and wishbone
Engine Performance		Rear	Semi-elliptic
Max. b.h.p.	75	Shock absorbers	Girling telescopic (front), piston (rear)
at	4,500 r.p.m.	**Dimensions**	
Max. b.m.e.p.	130	Wheel type	Steel disc
at	2,600 r.p.m.	Tyre size	5.50 x 15
B.h.p. per sq. in. piston area	2.24	Steering gear	Worm and nut
Peak piston speed ft. per min.	2,680	Steering wheel	16 ins. three-spoke
		Wheelbase	7 ft. 4 ins.
Engine Details		Track: Front	3 ft. 9 ins.
Carburetter	2 S.U.	Rear	3 ft. 9½ ins.
Ignition	Coil	Overall length	11 ft. 9 ins.
Plugs (make and type)	Champion L.10	Overall width	4 ft. 7¼ ins.
Fuel pump	AC mechanical	Overall height	4 ft. 3½ ins.
Fuel capacity	12 galls.	Ground clearance	6 ins.
Oil filter	Purolator by-pass	Turning circle	32 ft.
Oil capacity	13 pints	Dry weight	15¼ cwt.*
Cooling system	Pump and fan	**Performance Data**	
Water capacity	18 pints	Piston area, sq. in. per ton	43.9*
Electrical system	Lucas 12-volt	Brake lining area, sq. in. per ton	158.5*
Battery capacity	51 amp./hr.	Top gear m.p.h. per 1,000 r.p.m.	19.25
Transmission		Top gear m.p.h. at 2,500 ft./min. piston speed	80
Clutch	Borg and Beck 9 in.		
Gear ratios: Top	3.89		
3rd	5.15		
2nd	7.81	Litres per ton-mile, dry	4,100*
1st	13.15		
rev.	16.66		

* Maker's estimate and derivations therefrom.

THE MOTOR October 22, 1952

A New 2-litre Triumph Sports Car - Contd.

choke carburetter has been replaced by two semi-down-draught S.U. constant-vacuum carburetters, the peak of the power curve being thereby shifted from 4,200 up to 4,500 r.p.m., the peak of the torque curve having gone up from 2,000 r.p.m. (128 lb./sq. in., b.m.e.p.) to 2,600 r.p.m., at which point 130 lb./sq. in., b.m.e.p. is realised.

It will be seen that the maximum r.p.m. continues a low figure by modern sports-car standards and the maximum

BEING BUILT UP.—This close-up shows how the side members and frame are centrally braced, and also indicates the tubular body mounting.

EXPORT EMPHASIS.—The very high power/weight ratio and large displacement factor of the new Triumph should give it exceptional popularity in the export markets. The lines of the car are shown in this illustration of a left-hand drive model with, be it noted, a central remote-control gear lever.

piston speed is also well below 3,000 ft./min. The engine should therefore have inherent reliability and durability and the car should benefit from following the classic recipe of the largest possible engine in the smallest and lightest possible chassis.

A normal rear axle with hypoid bevel gears is used, the engine being connected thereto by a four-speed gearbox synchronized on 2nd, 3rd and 4th ratios, the estimated maximum speeds on the three indirect gears being 24, 45 and 70 m.p.h. respectively. A centre gear lever with remote control is provided.

As can be seen from an illustration, the frame consists of a simple structure with box-section side members and a distorted X-shaped centre bracing made from an open channel which connects four points to the side members by tubes which also act as body mountings. There are in addition a tubular rear cross-member, and a deep front cross-member which also gives support to the front suspension elements.

Concentric Springs and Dampers

These consist of fabricated wishbones connected to the front wheels through the medium of a ball-and-socket joint, the suspension itself being by open coil springs which surround direct-acting telescopic dampers. In order to increase the inertia of the front end of the frame structure a large-diameter tubular cross-member is used to join the top mounting points to each wishbone and yet further stiffness is added by a small diameter cross-tube at the extreme nose of the car.

Worm-and-nut steering is used in conjunction with a three-piece track rod and despite the short wheelbase of the car the front of the engine timing cover is not placed forward of the front wheel centres.

The rear axle is joined to the frame (which passes beneath it) through leaf springs which also drive the car, and these run downwards towards the front so as to promote an under-steering tendency as the car rolls. In view of the low centre of gravity it has not been thought necessary to provide an anti-roll bar and in order to have the lightest possible point of attachment the rear spring dampers are of piston type bolted on to short brackets welded on to the side members of the frame.

The wheels themselves are of the steel disc type with ventilating holes to assist in cooling the Lockheed-operated brakes which are of the two-leading-shoe type on the front wheels and largely in consequence of drum width of $1\frac{3}{4}$ ins., have a bigger than usual friction lining area.

It will be seen that in the general mechanical specification the car seeks usefully to employ known knowledge and components and to provide high performance by sensible proportions. The bodywork of the car has been designed with a view to economical production, rigidity, and moderate drag.

As can be seen from the illustration the nose of the car is sharply swept down to a rectangular, horizontal air inlet from which the air is ducted back towards a vertical radiator core. The bonnet has outside hinges attaching it to the scuttle and the separate front wings are bolted on and swept right through the doors to meet the rear panel.

INHERENT STABILITY.—In order to promote basic understeering qualities the rear spring leaves are inclined sharply downwards as shown in this picture, which also emphasizes the short, stiff propeller shaft.

THE MOTOR October 22, 1952

The body sides are very sharply cut away so as to provide ample elbow room for the driver and although a single-piece flat windscreen set at an angle of 40 degrees from the vertical is supplied as standard, it is quickly detachable and provision made for alternative aero-type screens. The rear wings are bolted on to the main structure of the body and can therefore be quickly replaced if damaged.

Two separate bucket seats are mounted within the cockpit, which has an interior width of 47 ins. These seats are of hammock type upholstered in leather and the body space behind them is entirely free for luggage, as the spare wheel is externally mounted in a recess formed in the back panel of the body.

In view of the fact that the whole of the engine and gearbox assembly is behind the front wheel centres, and the fuel tank and the contents (weighing up to 85 lb.), and spare wheel, are both behind the rear wheel centres, the car should have a distribution of weight as between the front and rear wheels appropriate to a high performance vehicle.

A three-spoke steering wheel is used and the instruments comprise a tachometer and speedometer each of 5 ins. diameter placed immediately in front of the driver, with a separate centre panel containing water temperature, petrol, and oil pressure gauges. There is also an ignition warning light and headlamp main-beam warning light. An open glove compartment is placed in the facia on the side remote from the driver, and refinements which are worthy of mention are the spring counter-balancing of the bonnet so that it may be quickly lifted, and the fuel filler cap placed within the spare wheel hub centre and fitted with a lock.

It will be seen that separate side lamps are mounted in the top of each front wing, the headlamps being semi-submerged in the front cowl. Semaphore-type direction indicators are not fitted but as an option the two tail lamps can be used as winking direction indicators (in countries where this is permissible).

This brief survey of the body specification shows that practical features have been given a good deal of thought and, although by no means aerodynamic in shape, the car should have a moderate coefficient of penetration which, in conjunction with its small size, should give a maximum road speed fully appropriate to the size of the engine and distinctly impressive in relation to the selling cost of the complete car, which is planned for production in the Spring of 1953.

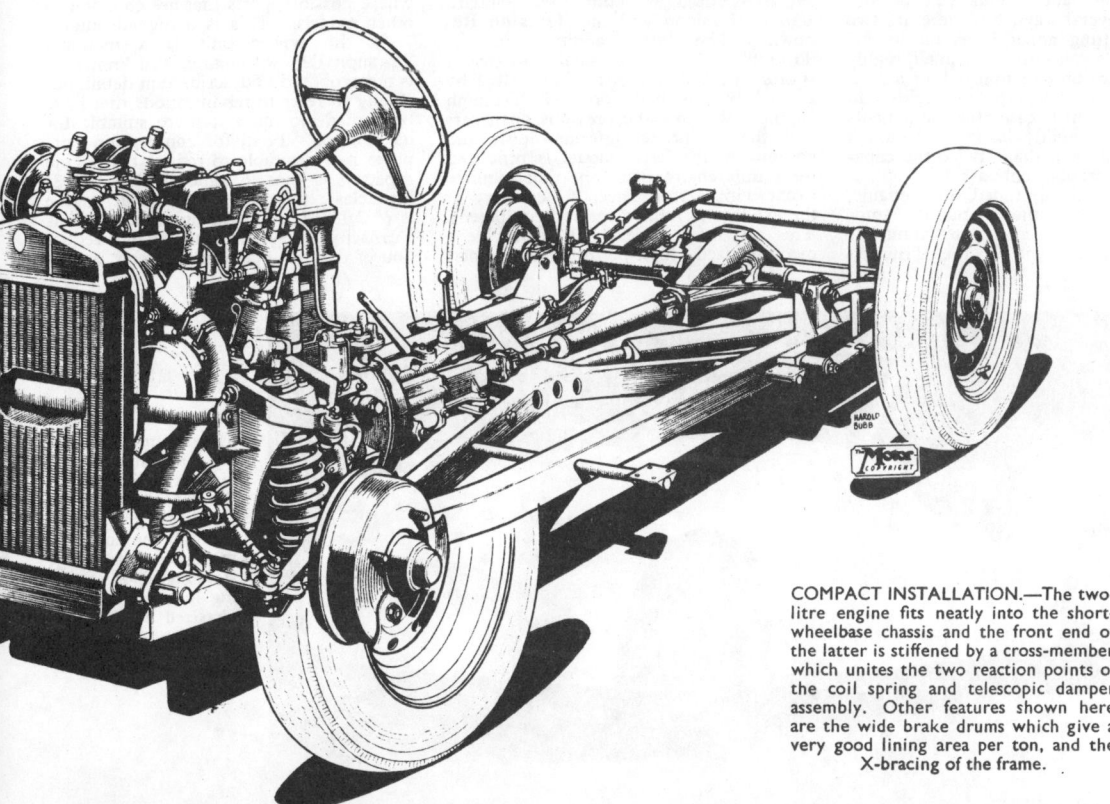

COMPACT INSTALLATION.—The two-litre engine fits neatly into the short-wheelbase chassis and the front end of the latter is stiffened by a cross-member which unites the two reaction points of the coil spring and telescopic damper assembly. Other features shown here are the wide brake drums which give a very good lining area per ton, and the X-bracing of the frame.

A TRIUMPH SPORTS MODEL

This new Triumph sports car has a neat and businesslike appearance, together with very sporting lines. The 2-litre engine and low overall weight should result in a model with a very good performance.

NEW CARS DESCRIBED

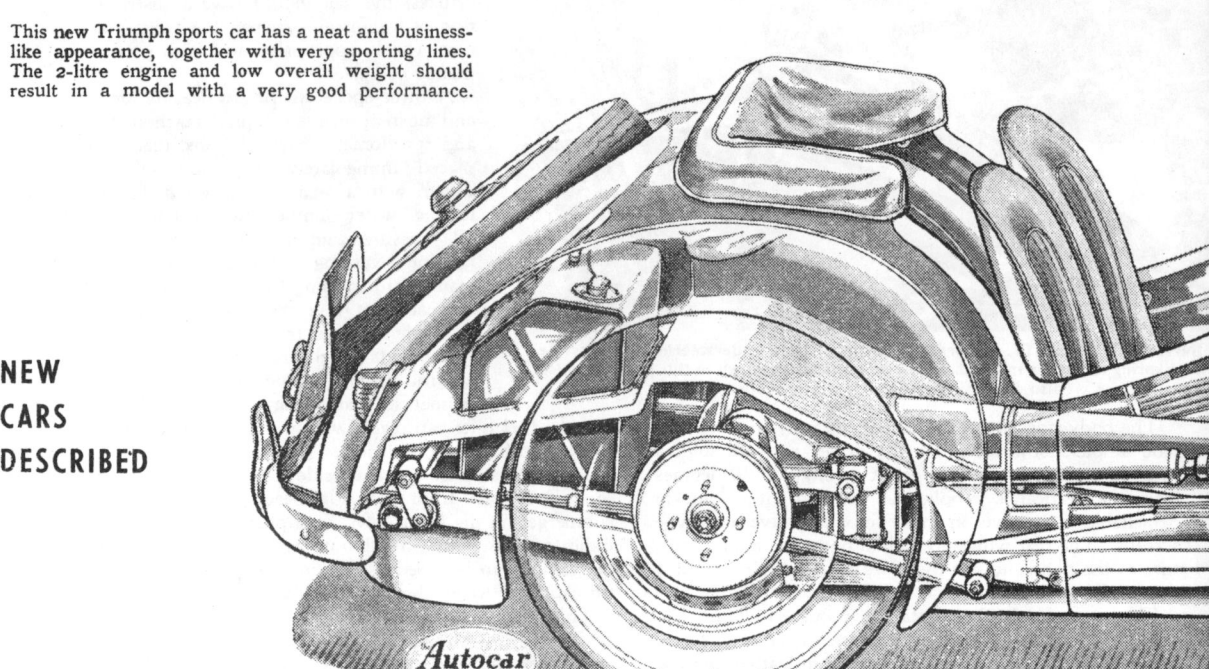

ONE of the essential requirements that must be fulfilled to produce a satisfactory sports car is a good power-weight ratio. This can be obtained in several ways, but there are two ways of setting about it as far as the power unit is concerned. A small, highly tuned engine can be employed, or a relatively large, touring type of engine can be used, in which case the components will be quite lightly loaded. Although the engine is of perhaps twice the capacity, it may produce only a similar output to that of the highly tuned small engine.

Since the war the Standard Motor company has produced one or two models of the roadster type in the Triumph range, the first of these being the 1800 Triumph, which was later modified to use the 2-litre overhead-valve power unit which is used in both the Standard Vanguard saloon and the Triumph Renown. The latest addition to the Triumph range is a small two-seater sports car, and this is also propelled by a slightly modified form of Triumph engine. Whereas the previous open cars had the accent on interior finish and equipment, the latest model is produced for a different market—one that requires a car with performance, if necessary at the expense of refinement and equipment. This car, then, is designed to provide a medium-sized high-performance sports car at relatively low cost. To keep costs down to an absolute minimum with limited production it is necessary to use, where possible, parts that are common to other models. This is a big advantage.

As the power unit is a modified Triumph Renown engine, well known, it is not proposed to describe it in detail, but chiefly to refer to modifications that have been made to make it more suitable for its new job. First, for competition purposes it is desirable to reduce the engine capacity so that it comes within the 2-litre class; the present Triumph engine is of 2,088 c.c., but the sports engine has a capacity of 1,991 c.c. obtained by a reduction of the cylinder bore diameter, a very

The Triumph sports car has simple but distinctive lines, with the front wing line extending back through the doors to meet the rear wing stoneguard. A sporting look is also emphasized by the cutaway doors.

2-LITRE TWO-SEATER, FOUR-SPEED GEAR BOX, 75 B.H.P. AT 4,300 R.P.M.

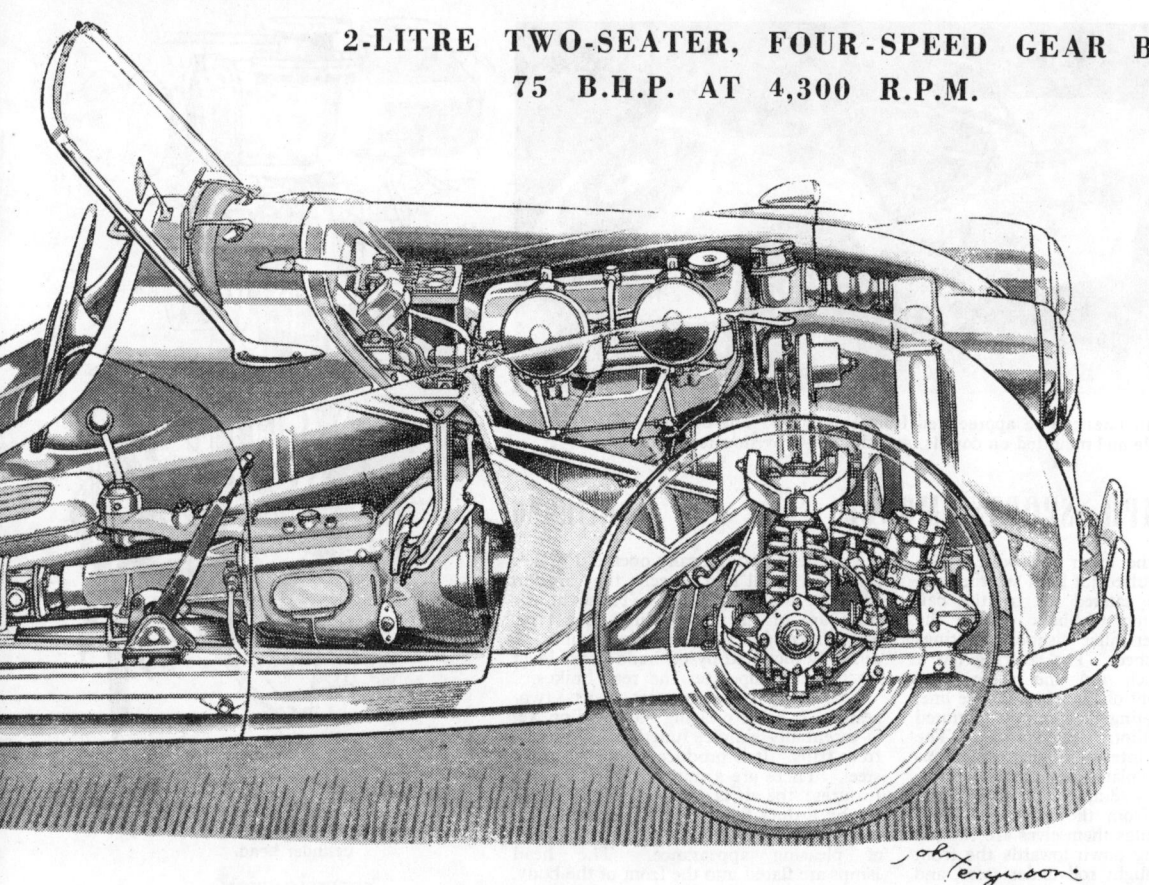

simple matter in an engine fitted with detachable wet cylinder liners, as it is necessary only to substitute liners with a smaller bore. The valve gear has been modified and uses a conventional arrangement of two springs per valve held in place by a collar and split collets. As used in previous models the engine uses a form of silent-valve gear whereby only the inner valve spring is connected to the valve by means of a collar and the outer valve spring is held in place by a cap that fits on top of the valve stem but is not attached to it by means of cotters. The camshaft is modified.

Fuel is supplied by twin carburettors on a special manifold, either semi-downdraught or downdraught type. Small circular silencers are fitted to the carburettor intakes. The exhaust system has a four-branch manifold.

To permit a low bonnet height several small modifications have been made to the front end of the engine. The cooling fan is mounted on and directly driven by the crankshaft; this operates in conjunction with a special low-mounted radiator. Consequently, the radiator top tank is not the highest part of the cooling system, and the system is filled through a cap attached to a casting on the engine. In other respects it is standard. With these modifications the engine develops 75 b.h.p. at 4,500 r.p.m., running with a compression ratio of 7 to 1.

Power is transmitted by a standard 9in nine-spring dry single-plate clutch to a four-speed gear box. The clutch is operated hydraulically by a compact unit mounted on the scuttle, which contains the master cylinders for both clutch and brake operation. These demand pendant pedals. The standard transmission used on both the Vanguard and the Triumph Renown is a three-speed gear box with synchromesh on all forward gears, but for this sports car the box has been redesigned, and, by providing synchromesh on top, third and second gears only, it has been possible to arrange four forward speeds in a box of a comparable size. Single helical gears produced from E.N.18 steel are used for all speeds fitted with synchromesh, but first and reverse gears are E.N.355. The gear box main shaft is E.N.352. The gear box layshaft runs on needle roller bearings fitted at each end. Gears are selected by a centrally placed remote control gear change lever.

Box Section Frame

A short propeller-shaft transmits power to the semi-floating hypoid rear axle, the casing of which is a very rigid structure and houses the taper roller bearings used to support the pinion. A two-pinion differential is used. Apart from the ratio and the track, the axle is similar to that used on the Triumph Mayflower.

To reduce overall weight as much as possible the vehicle is compact, and the power unit is mounted in a small box section chassis frame. This consists basically of two side members; the box section is formed by welding an angle section to an unequal channel section, in such a way that, apart from the box section so produced, there are also flanges projecting inwards and downwards to increase the depth of the structure and improve stiffness. An unequal cruciform is produced from channel section boxed at the centre. The rear is attached to the main frame side members at the attachment point of the rear spring front anchorage. The centre is drilled to enable the exhaust pipe to pass through it (owing to the low build of the car the transmission line runs above the frame). In front of the cruciform centre section is a small, detachable cross member on which is the rear power unit mounting. Tubular cross members are attached to the extreme front and rear of the frame (the rear one also forms the suspension anchorage points). The main front cross member is set some 10in back from the front of the frame side members, and there are also two built-up structures consisting of welded pressings to form the front suspension attachment points, the spring reaction member and bump stop, and the front engine mounting supports.

These structures are arranged on each side of the main cross member and are braced by a second tubular cross member that is detachable, and additional stiffening is provided by means of a strut attached to the top of the spring abutment and extending back to a point on the side member to form a triangulation. Its attachment point to the side members is where the front cruciform bracing members also join. The intermediate body attachment brackets are carried on short, tubular members, which extend through the frame side members and project through, and are welded to, the cruciform members.

The front suspension is independent by long and short wishbones, and is similar to that used on the Triumph Mayflower, except that stampings are employed for the lower wishbone members. The top wishbones are pressings. A housing for the ball-ended swivel pin is attached to the outer end of the top wishbone, and at the lower end the swivel pin is fitted in a screwed bush that is pivoted on the

The neat clean lines can be appreciated from this view. The curved windscreen is detachable and mounted on dowels, and there is provision for aero screens.

A TRIUMPH SPORTS MODEL . . . . . continued

outer end of the lower wishbone. The movements required for both steering and suspension are catered for by the ball joint on the top wishbone. Both the upper and lower inner wishbone bearings are rubber bushed. The steering has a three-piece track rod and slave lever mounted in front of the wheel centre line.

The half-floating rear axle is mounted on two half-elliptic leaf springs. The springs are of interest in that they contain only four plates but are 2in wide. Spring length is 38in, and the axle centre line is 17⅝in from the front anchorage point. The springs themselves are slightly inclined, sloping down towards the front to produce a slight roll under-steer, and the relatively wide spring dimension helps to prevent sideways movement of the axle.

Lockheed hydraulically operated brakes are fitted to all four wheels; they are 9in diameter and 1¾in wide. Those at the front are of the two-leading shoe type. They operate in cast iron alloy drums. A centrally mounted fly-off hand brake lever mechanically operates the rear brakes.

The body is a compact sports two-seater. It has a pleasing modern line, yet is essentially simple, functional, and free from frills. It is produced from 20-gauge steel. There are a number of interesting features: the air is conveyed to the radiator by a duct fitted with a stoneguard at its inner end; a new front end treatment of pleasing appearance. The head lamps are flared into the front of the body, and the front wing line runs right back through the doors, themselves low cut to provide ample arm clearance. Both rear wings are detachable, as also is the front pressing containing the head lamps and air intake, a feature that considerably facilitates servicing. The spare wheel is carried on the rear body panel and the fuel filler cap extends through the carrier, so that the tank is filled through the hub. This arrangement reduces the length of the tank filler pipe and avoids either a bulge or an extra trapdoor in the side.

Luggage Space Provided

With the object of reducing weight to a minimum, no external luggage locker is provided, but there is luggage space behind the passenger seat in the compartment also used for hood stowage. The neatly shaped hood is attached to the windscreen by a row of fasteners, and the screen itself is quickly detachable by undoing four fasteners. It is positioned on dovetails, and provision is also made for the substitution of aero screens.

For the body the accent has been on functional simplicity. The interior, including the facia panel, is trimmed with leather, the tops of the doors are padded with rubber, and there is a passenger grab handle attached to the scuttle—a very desirable feature for any fast car. The individual seats are both provided with adjustment, and on the central tunnel enclosing the gear box there is a cut-away portion, which forms a footrest for the driver's left foot—also very desirable.

The new Triumph is thus a medium-sized high-performance sports car that is functional and should meet the requirements of a very large number of motoring enthusiasts

To permit a low bonnet line the fan is mounted on a crankshaft extension and the radiator is filled through a casting attached to the top of the cylinder head.

SPECIFICATION

Engine.—4 cyl, 83 × 92 mm (1,991 c.c.). Compression ratio 7.0 to 1. 75 b.h.p. at 4,300 r.p.m. Torque 105 lb ft at 2,300 r.p.m. Three-bearing crankshaft. Inverted "bath-tub" combustion chambers. Side camshaft operating vertical overhead valves by push rods and rockers.

Clutch.—Borg and Beck dry single-plate, 9in diameter, nine-spring; hydraulically operated; carbon thrust withdrawal mechanism.

Gear Box.—Four forward speeds, synchromesh on top, third and second. Overall ratios: Top 3.89; third 5.15; second 7.81; first 13.15; reverse 16.66 to 1. Central remote control lever.

Final Drive.—Hypoid bevel. Ratio 3.89 to 1 (9 : 35). Half-floating axle shafts.

Suspension.—Front, independent by coil springs and wishbones; telescopic dampers. Rear, half-elliptic leaf springs; piston-type dampers and anti-roll bar. Suspension rate (at the wheel), front 82lb per in; rear 90lb per in. Static deflection, front 5.3in; rear 4.25in.

Brakes.—Lockheed hydraulically operated two-leading shoe front; leading and trailing rear. Drums 9in × 1¾in front and rear. Total lining area 121 sq in (60.5 sq in front).

Steering.—Worm and sleeve.

Wheels and Tyres.—Dunlop 5.50-15in on 4-stud 4.50-15in perforated steel disc wheels.

Electrical Equipment.—12-volt, 51 ampère-hour battery. Head lamps, double dip, 48-48 watt bulbs.

Fuel System.—12-gallon tank. Oil capacity 13 pints; by-pass filter.

Main Dimensions.—Wheelbase 7ft 4in. Track, front 3ft 9in; rear 3ft 9½in. Overall length 11ft 9in. Width 4ft 7½in. Height (in running trim with hood erect) 4ft 3in. Ground clearance 6in. Frontal area 15 sq ft approx. Turning circle 32ft. Weight (in running trim with 12 gallons fuel), 15¼cwt (1,708lb).

Price.—£555. With British purchase tax, £864 16s 8d.

A combined master cylinder unit is attached to the engine bulkhead, and its pendant pedals operate the clutch and the brakes.

May 27, 1953

THE MOTOR

The Birth of a SPORTS CAR

The Story of the Design and Development of the New Triumph Model

IT all began with a visit to the United States in the summer of 1952 by Standard's deputy chairman and managing director, Sir John Black. Object of his visit was to find out how his company could earn some dollars for Britain, and he soon came to the conclusion that the Vanguard was too like an American car to compete successfully with the home product. What was wanted was something distinctive, a type of car not made in the United States.

If only a sports car having a high performance and yet selling at a moderate price could be produced, Sir John felt that his company could enter the American market with every hope of success. When he returned home in July, therefore, Sir John asked his technical director, Mr. E. G. Grinham, to investigate the possibility of producing a sports car with traditional lines, capable of 90 m.p.h. and selling at a target price of approximately £500.

This target figure set the chief engineer, Mr. L. H. Dawtrey, no mean problem. It meant that the car would have to be built from Standard and Triumph components already in production in order to reduce tooling costs. It also meant a limited sum could be spent on tooling charges for the new body, a fact which eventually had considerable influence on the type of body fitted.

It did not take long to decide that the main components of the new car should consist of a Standard Vanguard engine and gearbox, and Triumph Mayflower front suspension and rear axle. Preliminary calculations soon showed, however, that even though the car was kept as compact as possible, not enough power would be available to propel a traditional type of car in outline at the set target speed of 90 m.p.h. owing to the high air resistance of this type of body.

The chief body engineer, Mr. W. J. Belgrove, therefore designed the present body with its back taper form to give the lowest possible drag in order that the car should prove capable of reaching its target speed and would also be economical on fuel at high cruising speeds.

While Mr. Belgrove was busy designing the body round a pair of side-members, the chief chassis engineer, Mr. H. G. Wenster, was hard at work modifying a Standard Eight chassis frame to accommodate the selected units. Drawings of the car were begun in late July, and it was only six weeks before Earls Court was due to open its doors that actual construction of the first prototype began, which meant that late nights became the standard routine for everyone working on the project, while week-ends as such ceased to have any meaning. Four sets of parts were put through, and from them three cars were built up.

The work went well, however, and no particular snags were encountered, although converting the three-speed

The Birth of a Sports Car

IMPROVED FORM.—The longer tail has not only improved the form of the body and so reduced its drag, it has also improved the weight distribution and has provided greatly increased room for luggage. The spare wheel now has a compartment of its own and the central filler for the petrol tank no longer projects through the hub of the spare wheel, as was the case on the prototype.

Vanguard gearbox to a four-speed box while retaining the same casting and making it on the same machine tools proved a very real headache, but even this problem was finally solved.

Such excellent progress was made that the first prototype was completed well before the Show opened and had done some running on the road before it appeared on the stand. The new car aroused tremendous interest at Earls Court, and in general it received a very favourable reception from the milling throng that practically submerged it every day. Some felt, however, that the compactness had, perhaps, been a trifle overdone, and a longer tail would improve the appearance of the car considerably, and would also allow more luggage to be carried.

When the Show closed its doors, the car went back to the engineering department and its evolution began in earnest. To assist in this development work, Ken Richardson joined the company as high-speed test driver on November 5, and at once took the car to the M.I.R.A. proving ground at Lindley to see how it handled. Richardson has probably had more experience than anyone else in the country of the exacting work of high-speed testing, for he carried out much of the testing of the E-type and other E.R.A.s and of the B.R.M. in its various forms.

Fast and No Fading

It was soon seen that although the Triumph Mayflower brakes had been replaced by Vanguard brakes with 9 in. by $1\frac{1}{4}$ in. drums, the braking still was not good enough for the speeds the car showed itself capable of attaining. The front of the body was therefore cut away slightly to direct cool air on to the drums, but only when 10 by $2\frac{1}{2}$ in. drums were fitted at the front did the braking become satisfactory. The bigger front brakes were first tried out in December of last year, and the correct size of drums having been found, further experiments were carried out with different linings. As a result of these tests, a fairly hard lining has now been adopted which seems immune from fade troubles however hard the car is driven.

The roadholding of the prototype was not all it might have been, and even the fitting of an anti-roll bar at the front did not greatly improve matters. It was decided that the modified Standard Eight chassis frame was at the bottom of the trouble owing to its lack of stiffness, and a new frame was therefore designed in December with greatly increased sections, more cross-members and the side-members farther apart, so that the body was no longer riding on a knife edge. The new frame is, in fact, more than three times as strong torsionally as the original frame, and when Ken Richardson tried the first car to be fitted with it in March of this year, he knew that their roadholding problems were practically solved. The new frame has straight side-members with no kick-up at front or rear, and as the exhaust pipe passes through the cruciform cross-member, the underneath of the car is exceptionally clean and the fitting of an undershield is therefore an easy matter.

Now that the roadholding had been tackled at source, as it were, attention could be concentrated on improving the suspension. Mr. Dawtrey has specialized in suspension problems for more than 30 years, and this long experience has left him gloomily convinced that suspension theory by no means inevitably works out in practice, and that only painstaking trial-and-error methods will eventually produce the right answer. The front suspension seemed satisfactory right from the start, but a wide variety of rear spring stiffnesses and shock-absorber settings were tried before Ken Richardson was satisfied with both the road-holding and the riding comfort of the car. The final result is a car that rolls hardly at all on corners despite no anti-roll bars being fitted, and yet gives a very comfortable ride.

Shortening the Tiller

While suspension testing was being carried out, development work was also pressed forward on the steering gear, and its ratio was progressively raised to its present figure of $2\frac{1}{4}$ turns from lock to lock.

- - - - - Contd.

IN AND OUT.—The airflow through the unusual shaped air intake has proved to be highly efficient and keeps the engine cool at high cruising speeds. Warm air is vented through louvres at the rear of the bonnet, which is of the rear hinged alligator pattern.

Some idea of the extensive nature of the development running carried out at Lindley is evident from the fact that on one occasion, driving the second prototype, Ken Richardson covered more than 480 miles in a single day, averaging 76.6 m.p.h. for this distance on the unbanked test circuit.

The development of the engine had by no means been neglected while all this chassis modification work was in progress, and during the past months the output has been progressively increased from 76 b.h.p. at 4,500 r.p.m. to 90 b.h.p. at 4,750 r.p.m. It may be remembered that the engine fitted to the prototype exhibited at Earls Court differed from the normal Vanguard engine in having twin S.U. carburetters, and valve gear of the taper cotter pattern, which is more suitable for high speeds than the silent pattern normally fitted.

Development restarted from this point by experiments with various camshaft timings and cam lifts until finally a higher-lift camshaft with somewhat wider timings was selected. Further modifications to the valves and valve gear followed, the size of the inlet valves being increased by 1/16 in. and the rockers were drilled to the top to improve the oil feed to them.

A more robust design of big-end was next decided on and the big-end bearings were changed from white metal lined shells to the indium lead bronze pattern. The compression ratio has been increased from 7 to 1 to 8.5 to 1. This has involved no change in combustion-chamber shape, for the Vanguard head was designed in the first place for premium fuels. The original exhaust manifold has been redesigned to give an easier flow for the gases, thereby getting the heat away from the induction manifold. Only modifications to the ignition system are a different advance curve for the distributor and the fitting of sparking plugs one grade cooler than those normally recommended for the Vanguard.

As the radiator header tank is below the level of the top of the cylinder head, a small casting on the head now carries the filler cap, the thermostat and the radiator thermometer connection. The fan is driven direct from the crankshaft, on which it is, in fact, mounted, but so efficient has the somewhat original air intake in the nose of the car proved that in general it runs too cool rather than too hot.

To see what would happen, the engine was run up to

SUSPENSION.—This view of the front suspension shows the forward-mounted steering gearbox, this being a left-hand-drive car, and the hydraulic shock absorber of the telescopic pattern mounted within the coil spring.

WET WEATHER EQUIPMENT.—Protection against the weather is exceptionally good for a car designed primarily for high performance, and the side screens have been tested at high speeds to ensure that they remain in place.

The Birth of a Sports Car - - Contd.

5,250 r.p.m., when, in fact, nothing happened.

Development work has also been continued on the body, and the new, longer tail not only enhances the look of the car, it also reduces still further the already low drag and has improved the weight distribution and therefore the handling.

At the speeds which the car is now reaching, the wind pressure tended to bend the flat windscreen, but giving it a slight bow has stopped this trouble. Incidentally, the windscreen can be removed completely by undoing four Dzus screws. As the actual speeds attained by the car correspond very closely with its theoretical performance calculated from its frontal area, power produced and drag coefficient, the body must be very efficient from the aerodynamic point of view.

The car's chrysalis period is now over and production will begin in earnest in July, although for the time being almost the entire output will be shipped overseas to those dollar markets for which the car has been designed.

There is no intention at present of undertaking a competition programme with the car, and no works entries in, for instance, the Alpine Trial, are at present contemplated. No doubt the car will, however, be seen in competitions in the hands of private owners, and its first appearance will be awaited with interest, for here is a most promising new contender for top honours in rallies and club races.

October 21, 1953

1954 Cars

Triumph

THE Triumph Mayflower having been withdrawn coincident with the introduction of the new Standard Eight, the range of this make consists now of two cars only, both fitted with a four-cylinder engine of approximately 2-litres capacity. Of these, the Renown is a well-established type which is unique in being the only series production model now obtainable with razor-edge coachwork. The lines of this four-door saloon body are unchanged from previous years and with such items as separately mounted headlamps and front mudguards offer a combination of elegance and practical merit which many continue to appreciate. Internally, the bodywork is notable for the use of Dunlopillo upholstery and folding armrests for both front and rear seats. The very large windows in conjunction with the narrow pillars give exceptional visibility.

Although listed as an optional extra a large proportion of the Renown models are now supplied with the Laycock-de Normanville overdrive providing four forward speeds. This makes it possible to climb a 1 in 10 gradient on direct top and to maintain 65 m.p.h. at 3,000 r.p.m. in overdrive.

During the past twelve months the other model in the range, the 2-litre sports, has been subject to intensive testing and a great deal of development work. The car shown at Earls Court this year is in consequence a very different vehicle from the one exhibited in 1952 although the dimensions are unchanged except for a longer tail and a lower windscreen. Now developing 90 h.p. the engine is coupled to a four-speed gearbox for which a fifth overdrive ratio is available. The frame has been substantially stiffened and there have been numerous detail changes to springs and dampers so that the road holding of the car is now of an extremely high order. Performance can be judged from a power : weight ratio of 100 h.p./ton and in competition tune and form, timed speeds in excess of 120 m.p.h. have been achieved.

Intensive development has brought considerable changes to the Triumph Sports since last year's Show.

Triumph Sports Achieves 124 m.p.h.

REMARKABLE SPEEDS FOR LOW BASIC PRICE

Chief test driver Ken Richardson, who drove the car for the speed runs, stands beside the Triumph Sports fitted with the special optional equipment.

THE SPEEDS IN DETAIL

	Westward. m.p.h.	Eastward. m.p.h.	Mean. m.p.h.
In High-speed Trim			
Kilom.	123.860	125.882	124.889
Mile	123.414	124.740	124.095
Hood Erected and Normal Screen (using overdrive)			
Kilom.	114.245	115.484	114.890
Mile	113.708	114.686	114.213
Hood Erected and Normal Screen (using direct top)			
Kilom.	106.065	111.958	108.959
Mile	105.664	111.455	108.499

From Harold Hastings. Jabbeke, May 20.

AN entirely new conception of cost in relation to speed is heralded by the performances recorded at Jabbeke this morning by a production prototype of the new 2-litre Triumph Sports. That fact is obvious when one relates the speeds achieved (which are set out in full on this page) to the basic price of only £555 asked for this model in standard touring trim.

I was present throughout the tests, which took place on the now familiar stretch of the Ostend-Ghent motor road near Jabbeke and were held under the auspices of the Royal Automobile Club of Belgium, which provided official timekeepers and subsequently measured the engine. I have witnessed various similar tests on this remarkably fine twin-track highway, but can recall none held under more perfect weather conditions; from early morning until the final runs were put in at about 11 a.m., the sun shone from a cloudless sky, the temperature remained moderate and what movement of air there was amounted to no more than an almost imperceptible drift.

The car was run in two forms—first in high-speed trim and then in more-nearly-standard form—and tests in the latter case were duplicated to obtain the speeds in both overdrive and direct top gears.

Variations from standard for high-speed trim consisted of removal of front and rear bumpers, removal of hood and side screens, together with the substitution of a small Perspex screen for the normal windscreens, the fitting of an undershield and rear-wheel spats, and the fitting of a rigid cockpit cover arranged to provide complete enclosure apart from a minimum-sized aperture for the driver.

High and Fast

In addition, a Laycock-de Normanville overdrive (which like most of the special equipment just enumerated, will be offered as an optional extra) was installed, this being solenoid-operated and arranged to give a 22% step-up, thus raising the 3.7-to-1 direct top gear of the four-speed gearbox to 3.034 to 1. In terms of r.p.m. and m.p.h. this raises the road speed at peak r.p.m. (4,800) from 96.5 m.p.h. to just over 118 m.p.h.

Other minor changes consisted of the obvious precautions of using Dunlop Road-Speed tyres in place of the standard type of the same size and the use of harder plugs (Champion L11S in place of L10S). Mr. L. H. Dawtrey, the chief engineer, told me that this plug change was the only point on which it was considered necessary to depart from the standard engine specification, the compression ratio being the normal 8.5 to 1 and the fuel Esso Extra.

The first run provided a paradoxically notable indication of the potency of this car. Growing from a tiny speck in the distance, it swept through the measured mile, obviously travelling very fast but sounding rather odd. The timekeepers announced a speed of 105 m.p.h. which seemed satisfactory enough to most of the spectators but did not please Sir John Black and those of us who knew what to expect. Ken Richardson (who drove throughout) coasted in shortly after to explain that he wanted a re-run. A plug lead had come adrift as he entered the timed section, leaving only three cylinders to complete the run.

After that, the car showed its true form—on which the figures can be left to speak for themselves.

In Standard Form

Subsequent to the runs in full high-speed trim, the cockpit cover was removed, and the normal full-width screen fitted and the hood and side screens erected to provide an indication of maximum speeds in more normal trim; in the time available, however, it was not possible to remove the undershield or fit the bumpers. Thus the figures recorded (see table) can be presumed to be slightly higher than those to be expected in touring trim, but are entirely consistent with the figure of 110 m.p.h. which the manufacturers are claiming as the maximum speed for this model in touring form.

Altogether a most interesting demonstration of a model which will be in full production within the next two months.

(An article on the development of this Triumph Sports will be found on pages 593-596.)

LATEST SPECIFICATION TRIUMPH SPORTS

ENGINE.—Dimensions: Cylinders, 4; bore, 83 mm.; stroke, 92 mm.; cubic capacity, 1,991 c.c.; piston area, 33.5 sq. in.; valves, overhead (push-rod); compression ratio, 8.5. **Performance:** Max. b.h.p., 90 at 4,800 r.p.m.; max. b.m.e.p., 145 lb./sq. in. at 3,000 r.p.m.; b.h.p. per sq. in. piston area, 2.69; peak piston speed ft. per min., 2,850. **Details:** Carburetter, two S.U.; ignition, coil; plugs (make and type), Champion L10S; fuel pump, A.C. Mechanical; fuel capacity, 12 gall.; oil filter (make, by-pass or full flow), Purolator by-pass; oil capacity, 13 pints; cooling system, pump, fan and thermostat; water capacity, 14 pints; electrical system, 12 volts; battery capacity, 51 amp./hr.
TRANSMISSION.—Clutch, 9-in. Borg and Beck; gear ratios: top, 3.7 (overdrive 3.034); 3rd, 4.9; 2nd, 7.4; 1st, 12.5; rev., 15.8; propeller shaft, Hardy Spicer; final drive, hypoid bevel.
CHASSIS DETAILS.—Brakes, Lockheed (2LS on front); brake drum diameter, 10 in. (front), 9 in. (rear); friction lining area, 148 sq. in.; suspension: front, independent (coil); rear, semi-elliptic; shock absorbers, hydraulic; wheel type, steel disc; tyre size, 5.50-15; steering gear, cam and lever; steering wheel, 17-in. spring-spoke.
DIMENSIONS.—Wheelbase, 7' 4"; track: front, 3' 9"; rear, 3' 9½"; overall length, 12' 4"; overall width, 4' 7½"; overall height, 4' 2" (hood up); 3' 10" (hood down); ground clearance, 6"; turning circle, 32'; dry weight, 17¾ cwt.
PERFORMANCE DATA.—Piston area, sq. in. per ton, 37.8; brake lining area, sq. in. per ton, 167; direct top gear m.p.h. per 1,000 r.p.m., 20.1; top gear m.p.h. at 2,500 ft./min. piston speed, 83.0; litres per ton-mile, dry, 3,350.

Sir John Black talking to Ken Richardson as he sits at the wheel of the Triumph sports car in speed trim.

THE

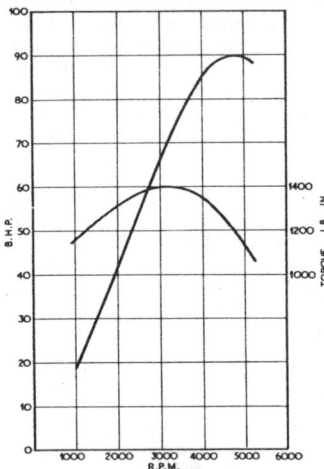

The Standard-built engine develops 90 b.h.p. at 4,800 r.p.m. running on first-grade pump fuel.

IT was recorded in the May 22 issue of *The Autocar* that the new Triumph sports car, driven by Ken Richardson, covered a kilometre at an average mean speed of 124.889 m.p.h. This in itself is no mean achievement for a two-seater sports car powered by an engine of under two litres capacity but, when it is realized that the car has a basic price of around £550, the result is outstanding. There are several cars with a performance that is comfortably over the magic three-figure mark, but Sir John Black has made it very difficult for the would-be purchaser of a sports car to find another car that would provide a comparable performance at anything like a comparable price.

The tests were performed in Belgium on the famous Jabbeke road, a highway of the *autobahn* type with twin tracks and flyover crossings. The road was officially closed and the runs were timed by the Belgian Royal Automobile Club. The complete set of results that were obtained is shown in the tables.

Weather conditions can change very considerably in a quite short space of time in the northern part of Belgium. On Monday, May 18, it was warm and dry;

on Tuesday it was cold with some rain and wind at times, and the glass was falling. Thus the big question of most people concerned was, what would the weather conditions be like on Wednesday? The quiet confidence of the technical director, Mr. E. G. Grinham, and his staff seemed to convey the impression that everything in *their* hands was well under control. In the early hours of Wednesday morning the team of technicians and mechanics assembled to have a quick cup of coffee before the journey out from Ostend to the Jabbeke road. The weather was kind; at 5 a.m. there was almost no wind and the sky was clear. As we drove towards Jabbeke there was a mist rising from the canals that line some of the roads, but by 6.30 this had cleared and visibility on the section of the road that was to be closed for the tests was perfect; wind speed was almost negligible, a mere 120ft per minute being recorded. By 7 a.m., when the gendarmes were at their posts to close the section used for the tests, the wind speed had increased slightly and was now 200ft per minute at an angle of about 60 degrees to the road. As this is equal to a speed of only just over 2 m.p.h. it was not particularly alarming, and it did not affect the averages very much.

As the neat form of the light-grey car sped down to the pits at the far end of the straight, ready for the start of the first run, the stage was set. Monsieur Lamot was at his station in the timing box and the trip cottons at the mile and kilometre positions were in place. The first run had started. As the car came past the measured section it seemed to be going quite fast but the engine did not sound as smooth as it had on previous occasions, and there were serious expressions on the faces of responsible people as they waited for the result to be calculated: 104.86 m.p.h. for the mile—not bad going for a 2-litre car but not good enough for the officials of the Standard Motor Company. A few minutes later Ken Richardson reported back to the control to tell the worst—a plug lead had slipped off as the car entered the timing section and the car had, in fact, attained a three-figure speed on three cylinders, or 1½ litres!

THE TRIUMPH TR2 2-LITRE SPORTS CAR

I. In speed trim using overdrive, with metal tonneau cover and undershield.

Distance	First run	Second run	Mean speed
1 Kilometre	199.335 k.p.h. 123.860 m.p.h.	202.588 k.p.h. 125.882 m.p.h.	201.005 k.p.h. 124.889 m.p.h.
1 Mile	198.615 k.p.h. 123.414 m.p.h.	200.749 k.p.h. 124.740 m.p.h.	199.711 k.p.h. 124.095 m.p.h.

II. In touring trim with hood up, undershield and using overdrive.

Distance	First run	Second run	Mean speed
1 Kilometre	185.854 k.p.h. 115.484 m.p.h.	183.861 k.p.h. 114.245 m.p.h.	184.889 k.p.h. 114.890 m.p.h.
1 Mile	184.569 k.p.h. 114.686 m.p.h.	182.995 k.p.h. 113.708 m.p.h.	183.807 k.p.h. 114.213 m.p.h.

III. In touring trim with hood up, undershield but without overdrive.

Distance	First run	Second run	Mean speed
1 Kilometre	180.180 k.p.h. 111.958 m.p.h.	170.697 k.p.h. 106.065 m.p.h.	175.353 k.p.h. 108.959 m.p.h.
1 Mile	179.369 k.p.h. 111.455 m.p.h.	170.050 k.p.h. 105.664 m.p.h.	174.611 k.p.h. 108.499 m.p.h.

THE AUTOCAR, JUNE 5, 1953

With this fault rectified the car toured down to the run-in for the start and a further test. With the news of a speed of over three figures on three cylinders it seemed logical to suppose that something really outstanding would be reached when it was running on four. We were not disappointed; the car sounded very well as it flashed past the group of spectators around the timing box, sounding very much like something jet propelled. For the measured mile the car recorded a mean speed of over 124 m.p.h. While breakfast was being served to the fortunate ones, there was much activity in the pit area where a well-trained team of mechanics removed the metal cockpit cover and converted the car back to normal touring trim, except that the undershield was still fitted and the bumpers were removed. In this trim, with hood and side screens erected, the car was only 10 m.p.h. slower, covering a mile at over

Watched by the Standard managing director, in the foreground, the car passes the timing caravan on the Jabbeke road and enters the measured mile.

PRICE of SPEED

AN OUTSTANDING PERFORMANCE BY THE TRIUMPH TR2 SPORTS CAR

114 m.p.h.—a very creditable performance.

What is this car really like, and how does it perform under ordinary conditions? In order to answer this question *The Autocar* drove the car after the photographers had taken their last shot of it on the Jabbeke road—in other words, in exactly the same trim as it was for the second and third set of tests but with the hood down, as the mid-day sun was comfortably warm. On the Jabbeke road it is not easy to get a general impression of how a car handles, but from the start the Triumph seemed right. It had a balanced feel coupled with a very definite amount of urge (there was marked acceleration at around 80 m.p.h. in overdrive) a quality resulting from a power output of 90 b.h.p. together with an all-up weight of 18 cwt. In spite of the fairly high compression ratio of 8.5 to 1, the car was very tractable at the bottom end and, on the first-grade pump fuel used for all tests, the engine was very smooth. Once off the motor road one goes from the sublime to the *pavé*, so in a short distance in Belgium it is possible to try a car over a large variety of conditions. For the high-speed runs the tyres were set at a pressure higher than that used for normal touring and they were not re-set before the car was tried on the rough surfaces. Nevertheless, over the *pavé* the ride was good; it was firm and well controlled without being harsh. The scuttle was also rigid and free from cross shake, an important point with an open two-seater car. On bends the roadholding was very good and the steering light and accurate; there was no suspicion of lost motion in the mechanism or spring in the steering rods. The car also had a pleasing amount of under-steer.

E. J. R.

Performance data
Maximum b.h.p., 90 at 4,800 r.p.m.
Maximum torque, 117.5 lb ft at 3,000 r.p.m.
Piston speed, 2,850ft per minute at 4,800 r.p.m. (equivalent to 100 m.p.h. in top gear).
Weight in touring trim complete with fuel, oil and water, 18¾ cwt.
Overall gear ratios: overdrive 3.03 to 1, top 3.7 to 1; third 4.9 to 1; second 7.4 to 1; first 12.5 to 1.
M.p.h. per 1,000 r.p.m. in top gear, 20.
Brakes, total lining area 148 sq in = 158 sq in per ton.

The car ready for test runs in touring trim. The men who performed the transformation scene are grouped behind the car; they are, from left to right, Mr. F. Smith, Mr. I. Walton, Mr. W. Vickers, Mr. R. Wilson, a Lucas representative, and Mr. J. Parkinson.

THE 120 M.P.H. TRIUMPH SPORTS

A notable addition to the range of high-performance cars

The new Triumph which, in touring trim (shown here), covered a measured mile at 114.213 m.p.h. The power unit is a modified twin-carburetter version of the 2-litre Standard Vanguard engine.

MANY enthusiasts hold the opinion that to be worthy of the title, a modern sports car must be capable of at least 100 m.p.h. and that its roadholding, brakes and steering should match its speed. Such cars today, in this or any other country, are few indeed.

On May 20, however, the Triumph Sports model, which made its bow in its present form at Geneva in March, gave ample proof that it has joined the select handful of cars with truly high performance when it was officially timed on the Jabbeke motor road by the Belgian Royal Automobile Club at 124.095 m.p.h. over the flying mile.

The driver on that occasion was Ken Richardson, who was intimately connected with the development of the B.R.M., and is now chief test driver to the Standard Motor Co., Ltd., manufacturers of the Triumph range.

During its trials, the car ran in both speed and touring trim, and with and without overdrive. Speed equipment included a metal tonneau cover which decked in the passenger's empty seat, a small aero screen and an undershield extending the whole length of the car. In touring form, the normal, full-width screen was used, with the hood erected.

A glance at the figures recorded is enlightening: in touring trim and with overdrive the car clocked 114.213 m.p.h.; in normal top gear, 108.499 m.p.h., thus pointing the lesson that the way to maximum speed lies in reducing frontal area and selecting the correct gear ratio.

The engine of the Triumph is a modified version of the Standard Vanguard. The bore has been reduced from 85 mm. to 83 mm. to bring the car into the 2-litre competition class with a total displacement of 1,991 c.c. The compression ratio has been raised to 8½ to 1, and with twin S.U. carburetters, the engine now develops 90 b.h.p. A four-speed gearbox, with a 3.70 axle ratio, is standard, but Laycock-de Normanville overdrive, as used in the Belgian test run, is an optional extra which will appeal to many drivers.

Although the twin-carburetter manifolding is the most obvious modification to the Vanguard engine it is by no means the only one. Other developments include a high-lift camshaft, enlarged inlet valves and a redesigned exhaust manifold giving improved gas flow. The big-ends are of more robust pattern and the bearing material is now indium lead bronze.

Suspension is independent at the front, using large helical springs, wishbones and telescopic shock absorbers. At the rear, the springs are wide semi-elliptics damped by piston-type

(*Below*) The layout of the cockpit: instruments, gear lever and handbrake are all placed where they are most needed. Note the shock absorbing trim on scuttle and door panels.

(*Above*) The redesigned tail unit, with separate spare wheel compartment, provides plenty of luggage space.

THE LIGHT CAR

shock absorbers. The brakes are Lockheed hydraulics, with 10-in. drums at the front, and two-leading shoe operation; 9-in. drums are fitted at the rear.

As will be seen in an accompanying photograph the Triumph chassis is extremely rigidly constructed from channel pressings with cruciform bracing at the centre. The exhaust pipe is carried through the cross-bracing, an arrange-

With the hood neatly folded out of sight, the Triumph has shapely—and practical—lines. The windscreen is quickly detachable.

SPECIFICATION IN BRIEF

Engine.—4 cyl., o.h.v., 83 mm. by 92 mm. (1,991 c.c.). 90 b.h.p. at 4,800 r.p.m. Compression ratio, 8½ to 1. "Wet" cylinder liners. Twin S.U. carburetters; A.C. mechanical fuel pump. Plugs, Champion L10S. Sump, 13 pts.

Transmission.—Borg and Beck 9-in. s.d.p. clutch; gears, 3.70, 4.90, 7.40 (synchromesh) and 12.50 to 1; reverse, 16.66 to 1. Final drive, hypoid bevel gears, semi-floating axle shafts.

General.—Suspension: front, independent (helical springs) with telescopic dampers; rear, semi-elliptics with piston-type dampers. Brakes, Lockheed 2LS. Steering, worm and sleeve. 12-volt, 51-amp-hr. battery and equipment. 12-gal. fuel tank. Tyres, 5.50 by 15.

Dimensions.—Overall length, 12 ft. 4 in.; width, 4 ft. 7½ in.; height (hood erect), 4 ft. 2 in. Clearance, 6 in. Turning circle, 32 ft. Wheelbase, 7 ft. 4 in. Track, front, 3 ft. 9 in., rear, 3 ft. 9½ in. Weight (dry), 17¾ cwt.

Price.—£787 7s. 6d. (Basic, £555.)

Manufacturers.—The Triumph Motor Co. (1945), Ltd., Coventry, England.

ment which makes the fitting of an undershield a relatively simple business. Patient and exhaustive high-speed testing has resulted in roadholding of a very high order; the car is free from roll, the suspension firm without being harsh.

A criticism of the Triumph as shown at Earls Court last year concerned the available luggage space. In the latest model the tail has been extensively redesigned so that the spare wheel is carried horizontally in a separate compartment beneath a boot which is spacious enough for the needs of two occupants. The layout of the cockpit is essentially practical, and in a high-speed car, that includes comfort. The leather-upholstered seats give full support on fast corners, the instruments are grouped where the driver can see them and there is plenty of legroom on each side of the prop. shaft tunnel.

At a basic price of £555 (and costing less than £800, including Purchase Tax), the Triumph Sports brings the delights of high-speed motoring within the reach of a very large potential market of enthusiasts.

The sturdy chassis of the Triumph Sports is built up from channel steel pressings, centrally cross-braced. Note also the massive helical springs of the front suspension and the two large S.U. carburetters: the engine develops 90 b.h.p.

JULY 1953

ROLL-FREE: (Left) The author cornering fast at Silverstone with the fascinating TR2 sports two-seater Triumph.

John Bolster
THE TR
World's Lowest-priced
Possessing Remarkable

THE Triumph TR2 is the most important new sports car which has been introduced for some time. First and foremost, it is easily the cheapest genuine 100 m.p.h. car on the market, and it brings this performance, with acceleration and roadholding to match, within the reach of the man of moderate means for the first time. Secondly, its excellent weather protection, large luggage space, and good traffic manners, render it entirely suitable for shopping and going to work.

The basis of the TR2 is a low and rigid box-section frame, with cruciform bracing. It passes beneath the hypoid rear axle, from which it is suspended by underslung semi-elliptic leaf springs. These are shackled at their rear ends and inclined downwards to the forward anchorages, to give an understeering tendency. In front, there are unequal length wishbones with helical springs, which embrace telescopic dampers. The cam and lever steering box operates a three-piece track rod and slave arm, ahead of the wheel centres.

Bearing little resemblance, especially in performance, to the touring unit from which it was derived, the four-cylinder 2-litre engine develops no less than 90 b.h.p. at the moderate speed of 4,800 r.p.m. It has a combined block and crankcase with replaceable wet liners, and the counterbalanced crankshaft runs in three bearings. The duplex chain-driven camshaft operates the overhead valves through pushrods and rockers, while two SU semi-downdraught carburetters, with A.C. air filter-silencers, supply the mixture.

A Borg and Beck clutch drives a four-speed gearbox, with synchromesh on the upper three ratios. The box has an extension to the rear, which permits the use of a very short propeller shaft. The gearchange is operated by a short remote-control lever, and the optional overdrive unit, as fitted to the test car, may be brought into action at the touch of a switch.

The hydraulic brakes, as the data panel shows, have larger drums in front than behind, and the "fly-off" hand lever is a most welcome feature. A 12½ gallon petrol tank is mounted behind the seats, ahead of the luggage boot, to give correct weight distribution.

Considering the moderate overall dimensions of the car, the body is remarkably roomy. There is considerable parcel space behind the seats, and the boot is of generous size. The spare wheel lies flat in a

SPACIOUS: (Above) Generous luggage accommodation is a feature of the TR2. The spare wheel is carried in a separate compartment below the boot.

separate drawer, and cannot therefore scratch or soil one's personal impedimenta. The general appearance is neat and functional, while the performance figures prove that the shape is efficient aerodynamically. The outer mudguard panels are clearly arranged for easy replacement in the event of damage, and the lamps are mounted inboard of this vulnerable area.

On taking the wheel, one immediately feels at home. The bucket seats are at just the right angle, and give lateral support for cornering. The pedals are properly arranged for "heel and toe", and there is plenty of room to rest the left foot.

Tests—
TRIUMPH TR2

100 m.p.h. Sports Car is Revealed as ... cceleration and Superb Road-holding

New Road Test Series   No. 1

A short travel and light movement render the gear lever pleasant to operate. The clutch is fluent in action, and will stand up to repeated racing starts and snap gearchanges.

Once on the move, it is obvious that the acceleration is quite out of the ordinary, as the data panel and graph show. The engine is very smooth, and except at tick-over speeds feels more like a "six" than a high-compression "four". It has plenty of punch in the lower ranges, so that one can drive largely in top gear if so inclined. Mechanically, it is quiet, but the exhaust is a little

quick, and the gears are commendably silent. The well-chosen ratios give 50 m.p.h. on second and 80 m.p.h. on third gear at 5,000 r.p.m. On top, 100 m.p.h. represents only 4,800 r.p.m., which entails a piston speed of 2,850 ft./min. It will thus be seen that an overdrive is certainly not a necessity, though the luxury of cruising at 80 m.p.h. with the rev. counter only just over 3,000 r.p.m., is probably worth the extra price. The maximum speed is almost identical on direct top and overdrive, and the mean of runs in both directions gave me 103 and 104 m.p.h. respectively.

I am delighted to say that no automatic nonsense is fitted to the overdrive, and one simply moves a switch to put it in or out of action. The change may be made on full bore without shock, and the engine copes manfully with the very high ratio of 3.03 to 1, which gives some 26 m.p.h. per 1,000 r.p.m. Natur-

ally, the acceleration is noticeably less brisk, and it takes about five seconds longer to go from 80 to 90 m.p.h. When travelling fast, I usually switched on the overdrive at about 95 m.p.h. The car fairly flies up main road hills on normal top; for example, it exceeded 90 m.p.h. up Wrotham Hill before I had to shut off for the traffic lights.

A car of such notable performance naturally requires roadholding to match. Most small sports models require a good deal of holding at high speeds, and as the wheelbase of the Triumph is only 7 ft. 4 ins., I expected that it would become somewhat lively at three-figure velocities. In fact, nothing could be further from the case, and the machine runs straight and true at maximum speed with the driver's hands resting lightly on the wheel. The suspension is fairly firm, but the ride is quite comfortable.

These suspension characteristics, allied to a low centre of gravity, ensure entirely roll-free cornering— a most unusual virtue these days. The light, high-geared steering gives a good sense of control, and very fast cornering results only in a gentle four-wheel slide. Rear end breakaway may be finally provoked only if one takes leave of one's senses and enters a curve at a virtually impossible speed.

As the TR2 will certainly be raced by some owners, I drove it round Brands Hatch and Silverstone. It handled admirably at both tracks, and I greatly enjoyed the experience. The short club circuit

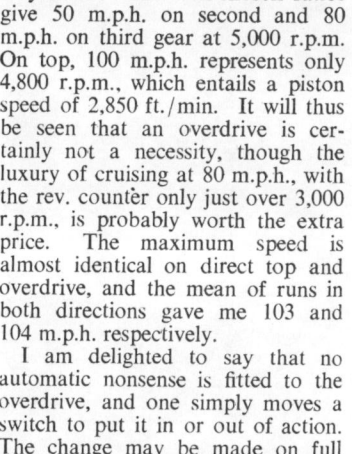

POWER-PRODUCER: (Above) The 2-litre push-rod-operated o.h.v. engine develops 90 b.h.p. on normal pump fuel. Air silencers are fitted to the dual S.U. carburetters.

on the loud side for my personal taste. It is not unreasonably noisy at the higher revolutions, but a resonance around 2,400 r.p.m. spoils one's silent passage through urban areas. In any case, some people still prefer a sports car to have a deep note, and this is a matter that can easily be altered to the owner's preference.

The gearbox earns absolutely full marks. The change is simple and

WEATHERPROOF: (Below) First-rate all-weather equipment is provided by the neat, plastic material top and close-fitting side-screens.

The TR2 has a maximum speed of well over 100 m.p.h., and John Bolster is seen here on the fastest section of the club circuit at Silverstone during a timed lap.

at Silverstone was slightly damp, and the large screen was erect, but I was able to lap in 1 min. 30⅛ secs., or 64 m.p.h. Under better weather conditions, and with only the optional aero screen in place, one could easily beat this figure.

The brakes are powerful, and entirely adequate for normal road work. Under racing conditions, however, they become very hot, and some additional air cooling might be provided for this work. Wire wheels are catalogued as an extra, and these would certainly aid heat dissipation.

It is a good point that all the instruments are separate, and have circular dials. The speedometer is only 2 m.p.h. fast, and a number of careful stopwatch checks proved that this small error does not increase at the higher readings. The general finish and interior furnishing are good, and certainly compare favourably with those of much more expensive machines. The lights were sufficiently powerful to allow me to drive at 100 m.p.h. in the dark.

The all-weather equipment is very good indeed. There are no draughts, and the visibility in all directions is excellent. The hood does not flap, and in fact I took the performance figures with the car closed. The sidescreens fit into very large sockets, and are unusually rigid. I occasionally bumped my knee on the forward anchorage, and a little extra padding here might be of value. Ease of entry and exit has

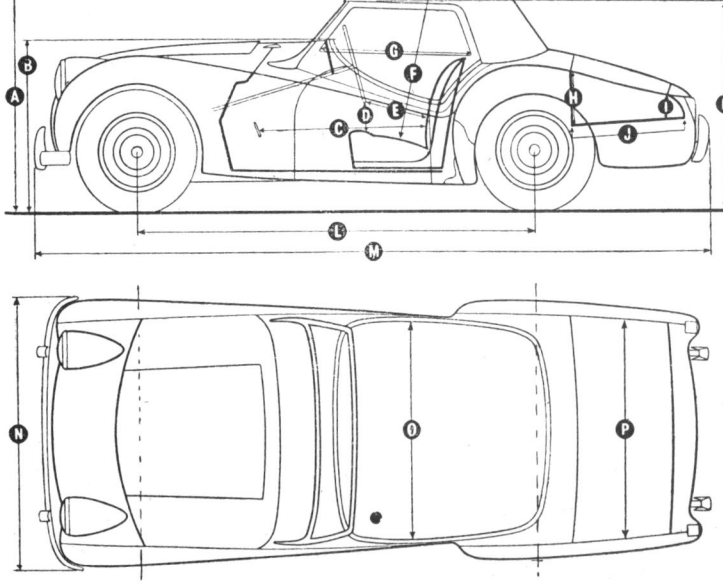

TR2 Dimensions

A. *Overall height, hood erect—4 ft. 2 ins.*
B. *Height of scuttle—3 ft. 4 ins.*
C. *Clutch pedal to seat squab—Max. 3 ft. 6½ ins. Min. 2 ft. 11 ins.*
D. *Steering wheel to seat cushion—6½ ins.*
E. *Squab to steering wheel—Max. 1 ft. 6 ins., Min. 9 ins.*
F. *Seat to hood when erected—2 ft. 11½ ins.*
G. *Window width at sill—2 ft. 7½ ins.*
H. *Height of Boot—Max. 1 ft. 2½ ins.*
I. *Height of Boot—Min. 7 ins.*
J. *Length of Boot opening—1 ft. 6¾ ins., Max. 2 ft. 2¾ ins.*
K. *Height to top of screen, hood folded—3 ft. 10 ins.*
L. *Wheelbase—7 ft. 4 ins.*
M. *Overall length—12 ft. 7 ins.*
N. *Overall width over bumpers—4 ft. 7½ ins.*
O. *Width at elbows—3 ft. 9 ins.*
P. *Width of Boot opening—3 ft. 5½ ins.; Max. width—3 ft. 9 ins.*

TRIUMPH TR2 – ACCELERATION GRAPH

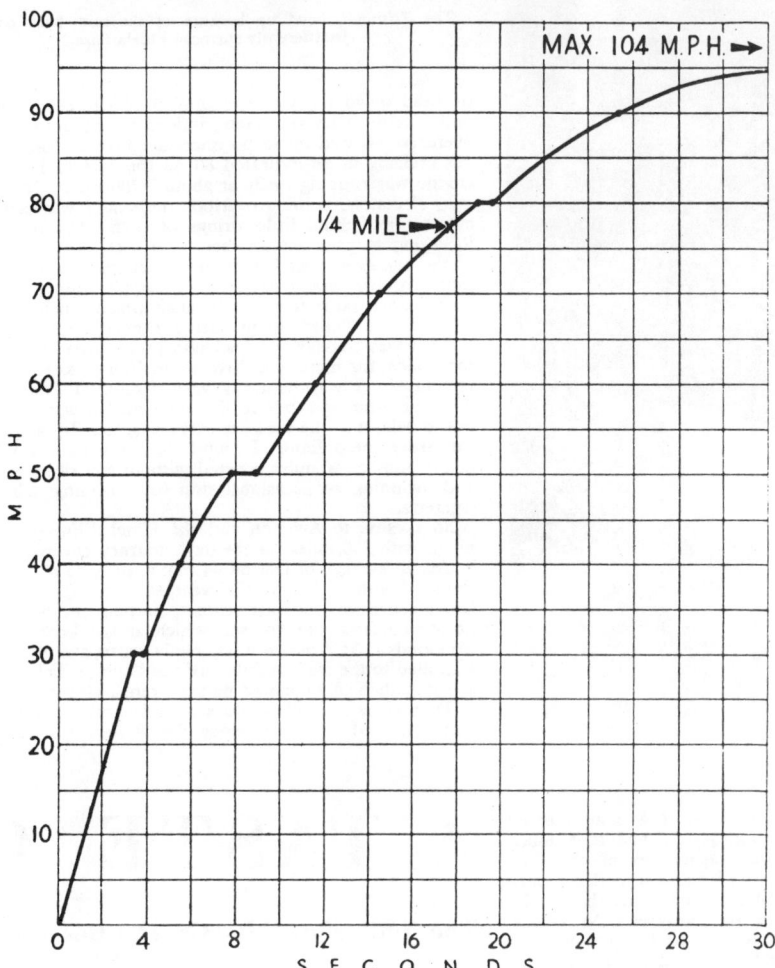

apparently been studied, for the doors can be negotiated without trouble, in spite of the very low build. In cold, wet weather I habitually drove without an overcoat.

Naturally, this is an ideal car for achieving high averages on our inadequate roads. The small overall dimensions are a great help in heavy traffic, and a touch of throttle sends the speedometer soaring round the dial whenever a clear stretch presents itself. The high-geared steering makes it easy to avoid the common clot, whether afoot or awheel, and emphasizes, once again, that a good sports car can be fundamentally a safer vehicle than a stodgy family saloon.

In spite of pressing on really hard, and exceeding 100 m.p.h. more times than I can remember, I achieved the excellent overall average of 25.4 m.p.g. If one were content to drive with a little less dash, a full 30 m.p.g. would readily be available. The Triumph is thus a cheap car to run as well as to buy. The low piston speed should give long wearing qualities, and accessibility is unusually good, which will appeal to those owners who prefer to carry out their own maintenance.

I am sure that the Triumph TR2 will meet the needs of many sports car drivers, and that this will become a very popular car. If you see that low, blunt nose in your mirror, pull over and let him go, unless you have something *very* hot!

Specification and Performance Data

Car Tested: Triumph TR2 Sports 2-seater, price £595 (£844 0s. 10d. with P.T.). Extra charge for overdrive, price £40 (£56 13s. 4d. with P.T.).

Engine: Four cylinders 83 mm. x 92 mm. (1,991 c.c.). Pushrod operated overhead valves. 90 b.h.p. at 4,800 r.p.m. 8.5 to 1 compression ratio. Twin S.U. carburetters. Lucas coil and distributor.

Transmission: Borg and Beck 9 ins. single dry plate clutch with hydraulic operation. 4-speed gearbox with short central remote control lever, plus electrically operated overdrive (optional extra). Ratios 3.03 (overdrive), 3.7, 4.9, 7.4 and 12.5 to 1. Short open Hardy Spicer propeller shaft. Hypoid rear axle.

Chassis: Box section frame with cruciform, underslung at rear. Independent front suspension by wishbones and helical springs with telescopic dampers. Cam and lever steering, 3-piece track rod. Semi-elliptic rear springs with piston-type dampers. Pierced disc wheels, fitted 5.50-15 ins. tyres. Lockheed hydraulic brakes, front 10 ins. x 2¼ ins. 2 L.S., rear 9 ins. x 1¾ ins. Total lining area 148 sq. ins.

Equipment: 12-volt lighting and starting. Speedometer, rev. counter, ammeter, water temperature, oil pressure, and fuel gauges. Flashing direction indicators.

Dimensions, etc.: Wheelbase 7 ft. 4 ins. Track, front 3 ft. 9 ins., rear 3 ft. 9½ ins. Ground clearance 6 ins. Turning circle 32 ft. Weight (kerb) 18¼ cwt.

Performance: Maximum speed (overdrive) 104 m.p.h. Speeds in gears: direct top 103 m.p.h., 3rd 80 m.p.h., 2nd 50 m.p.h., 1st 30 m.p.h. Standing quarter mile, 17.9 secs. Acceleration: 0-30 m.p.h., 3.4 secs.; 0-40 m.p.h., 5.5 secs.; 0-50 m.p.h., 7.9 secs.; 0-60 m.p.h., 11.6 secs.; 0-70 m.p.h., 14.5 secs.; 0-80 m.p.h., 19 secs.; 0-90 m.p.h., 25.6 secs.

Fuel Consumption: Driven hard, 25.4 m.p.g.

* * *

TULIP RALLY

THE 6th Tulip Rally, organized by the R.A.C.-West, takes place from 25th April to 1st May. Cars eligible are:— (a) Production touring cars; (b) Production "Gran Turismo" machines; (c) Modified production cars; (d) Production sports cars; and (e) International sports cars (conforming to Appendix C of the Sporting Code).

Entries close on 31st March at normal fees, i.e., 245 guilders (approximately £25) plus about £3 for insurance; all other passengers, other than named co-drivers, about £5 each extra. Closing date at approximately £7 extra, 12th April.

Starting points are from London, Berne, Brussels, The Hague, Hamburg, Munich and Paris. All routes converge on Nürburgring where a special test will be held. Thereafter all cars will follow a 775-mile route to the finish at Noordwijk. Average speed to Nürburgring will be 31 m.p.h., and thenceforth 34 m.p.h. However, higher averages may or may not be imposed for special stages included in mountainous country. All eliminating tests en route will be on roads closed to normal traffic.

The rally terminates with speed tests on Zandvoort circuit. The famous Tulip Ball takes place on 1st May.

As last year, classification is by groups, and the best overall winner in his (or her) group—i.e., having the biggest percentage margin over others in the class—will be the winner of the rally.

The event counts towards the European Grand Touring Championship, for which only standard touring cars are eligible.

The Triumph, hood up because of the rain, halts on the indifferently surfaced Flüela Pass.

its head down the vast stretches of N.6 through Tournus and Mâcon to Lyons. Here, indeed the advantages of the overdrive showed up in no uncertain manner for, with the car cruising at an indicated 80 on the 3.03 to 1 ratio, the engine was running easily at about 3,200 r.p.m. At such a gait, of course, nothing overtook me, and I was constantly coming up behind little strings of traffic, five or six cars jockeying to pass one another. On such occasions a flick of the overdrive switch into the direct top gear and a little more throttle would send the Triumph scampering past, with a crisp rattle from the exhaust which quietened down again as one flicked the overdrive in once more.

As it was Sunday the road was pretty busy and so were the police, for there is a drive on in France to reduce road casualties. I was naturally very careful at every village with the limit sign, but on the open road it was impossible not to delight in the car's capacity for speed. In fact, despite the many speed limits, I found that each hour would see nearly another 60 miles covered without any effort to do so, and including an occasional stop for a cooling drink or a cigarette.

So I came to Avignon and the Hotel d'Europe for the night, with 350 miles for the day's journey covered in eight hours, an average of just on 44 m.p.h. in spite of numerous stops. South of Lyons the weather had improved and I had taken down the side screens, keeping the hood up as protection from the hot sun which at last broke through the clouds. At Avignon it was really warm, and after dinner I strolled to the ancient Palace of the Popes, which has been so sympathetically restored by the French Government.

Then came a couple of days in real sunshine as the TR2 and I headed eastwards along the Riviera towards Italy.

A TASTE OF

The TR2 Shows Its Capabilities Over 2,40

As an observer at the Scottish Rally at Whitsuntide and at the Alpine Rally during July, I had been much intrigued by the obvious capabilities of the little Triumph TR2. When the Liège-Rome-Liège Rally came along, which I wanted to see, it occurred to me that a TR2 would be an ideal mount for the purpose. On Friday, August 13, which some might consider ominous, I collected a geranium-red TR2 at Coventry—actually the identical car of *The Autocar* Road Test early this year—and headed south for Dover on one of the few fine days of this appalling summer. Next morning I drove aboard the *Lord Warden*, with the sole object of getting as far south as possible.

It was a bright sunny morning as we docked at Boulogne, where the band of the Grenadiers and a French guard of honour were drawn up. A delicate compliment, I thought, but then found that the Mayor of Dover was visiting M. le Maire of Boulogne, and that the town was en fête. As a result, the normal route from the dock was obstructed, so that I had to plunge into the maelstrom of one-way streets in the town itself. When I passed one jovial *agent de police* for the second time he gave me a whimsical look and exclaimed "*Tiens! Encore?*" His instruction to turn left, and left again, landed me in a press of buses, cars and horse-drawn vehicles firmly wedged behind a road barrier. It was all very amusing, but occasioned a loss of valuable time.

At St. Pol I stopped for lunch and by the time I reached Soissons it was well into the afternoon, the sun had gone in and it began to rain. Night was falling—as well as heavy rain—as I ran into Troyes with only 250 miles on the clock. It was still raining next morning as I headed the little red car southwards along N.71, following the picturesque valley of the Seine to Châtillon and Dijon, but I remained snug and dry with hood and side curtains erected. I could not bring myself to hurry over the road from Dijon through Beaune to Châlon-sur-Saône, because the famous Burgundy vineyards were looking a picture, and did not seem to have suffered unduly from the recent storms.

To make up for my loitering, however, I gave the TR2

17 SEPTEMBER 1954

At Ventimiglia there was a long queue of cars three abreast, shepherded by half-a-dozen *agents*, waiting to pass through the frontier post. For one hour and forty minutes we sat and baked in the blazing sunshine, moving forward a few feet at a time. At last it was my turn to hand out my carnet and passport, which were quickly stamped, and then I continued my journey, thankful for the cooling breeze which movement of the car brought.

Along the winding road of the Italian Riviera it was a question of keeping in the queue, except when a brief straight and the TR2's vivid acceleration allowed one to jump to a gap ahead. The tourist traffic seemed astonishingly heavy with cars from Great Britain, Germany, Belgium, Switzerland, Sweden, Denmark and Morocco, but, of course, August is the peak of the holiday season on the Continent as in England. Not until I reached Genoa and turned northwards along the autostrada towards Milan was it possible to make any swift progress.

On the run to Milan, however, the overdrive again came into constant use and the engine seemed to improve in smoothness with a little full throttle burst, for along the slow winding road of the Italian Riviera it had evinced an occasional misfiring which I put down to dirty sparking plugs. The same tendency was noticed again on the winding road along the shore of Lake Como, but then, on the fast road through Sondrio to Tirano and Bormio, the engine again settled down to its steady beat.

Rain had now set in again, and I was sorry for the competitors in the Liège-Rome-Liège, who must have had an uncomfortable night's drive before, in the early morning, they climbed the Stelvio and ran down to the Bormio control, where I was awaiting them. Then came a hard day's motoring for the TR2, for I whisked it up the Stelvio, down to St. Maria, over the Fuorn Pass to Zernez and Sus, then over the Flüela to Davos, down the valley of the Landwasser to Tiefencastel and Thusis, swinging north to Bonaduz and turning westwards for Ilanz and up the valley of the Reno, which at Truns was a muddy flood threatening to overflow its banks. As a matter of fact, a few hours later it did, and there were four feet of water across the road.

But by then the TR2 had stormed over the Oberalp Pass, in rain which was almost solid, down into Andermatt where it was getting dusk. I was tempted to stay in comfort at Andermatt, but was due to make a rendezvous with an aircraft at Geneva next day. The muddy torrent of the river which flows down from the Furka was obviously threatening the road and the bridges, and it was already over the railway line in places. So, with some misgivings, for the clouds were low and made visibility so poor that lights were desirable, I pressed the Triumph on up the Furka.

Some of the passes I had already been over were decidedly loose and rough, and with so much power available in such a light vehicle I had found that a certain amount of care was necessary to retain maximum adhesion. However, the Triumph took me safely and swiftly to the top, the thermometer not having risen a fraction and the oil pressure gauge not having fallen in the least.

Roads Flooded

Then past the Rhône glacier down to Gletsch, to find that the run down the valley to Munster is now almost a boulevard. At Brigue I felt that both the TR2 and I had earned a night's rest, for the rain was still coming down in stairrods. In the morning mine host told me that the road I had travelled over the previous evening was impassable through floods.

As I ran into Montreux the rain ceased and the sun came through, and at Rolle, when I stopped for lunch, I looked out across Lake Leman, and could see the muddy waters of the Rhône making a wide brown strip down the centre of the lake. I got away from Geneva about 4 p.m. and the Col de la Faucille, with its good surface and easy gradient, was nothing to the TR2 after the rough and narrow mountain passes which it had already covered.

As I had to be alongside the ship at Boulogne next day, I had to "press on regardless" northwards, but the road through Morez, Poligny, Dole, Gray, where N.67 was followed to St. Dizier, then N.4 to Vitry-le-François, and then along N.44 to Chalons-sur-Marne and Rheims, is fast and I could use the overdrive to advantage.

420 Miles in a Day

Stopping for petrol about 100 miles south of Rheims, I rang up the Lion d'Or in that city to reserve a room, and said that I should arrive in two hours' time. It was now dark, but with yellow bulbs in the head lamps, which cut down the light a little, I had no flashing from other drivers and managed to complete the journey on time. This was the longest day's run, with 420 miles on the clock, but it was done so easily that in spite of numerous stops and delays it occupied a little less than twelve hours.

It was still dull next morning, but the 178 miles to Boulogne were an easy four-hour run. And so back to England in the *Lord Warden* once again. In the ten days the little TR2 had covered 2,400 Continental miles, including several of the highest Alpine passes and mostly in torrential rain. Hood and side curtains were kept up all the time except for the two days along the Riviera. Even in the worst rain I remained snug and dry, and on the highest mountain climbs the heater was utilized, for there was still quite a lot of snow about.

Of the car's performance there is little to say that has not already been said in *The Autocar* Road Test. Over the whole distance the fuel consumption worked out at 32 m.p.g., despite a high average speed from point to point and much third and second gear climbing of some of the highest mountain roads in Europe. The taste of the Triumph has certainly whetted my appetite for more.

TRIUMPH

Continental Miles By A. G. DOUGLAS CLEASE

Even in wet, cloudy weather the Alpine passes have a splendour of their own—as this photograph of the Fuorn shows.

26 JUNE 1954

TRIUMPH at Le Mans

A stock model Triumph T.R.2 Sports Car, privately owned and driven by E. B. Wadsworth and R. Dickson in the Le Mans 24-hour race.

1. Distance covered 1,804 miles
2. Average speed 74.71 m.p.h. for 24 hours
3. PETROL CONSUMPTION 34.688 m.p.g.

58 Starters TRIUMPH T.R.2 finished 15th

Price: £625 (P.T. £261 10s. 10d.)

TRIUMPH T.R.2 SPORTS

Triumph Motor Company (1945) Ltd.
SUBSIDIARY OF THE STANDARD MOTOR COMPANY LTD., COVENTRY

TRIUMPH CARS · STANDARD CARS · STANDARD COMMERCIAL VEHICLES · FERGUSON TRACTORS

October 20, 1954

1955 CARS

TRIUMPH

Detachable Hard Top Now Offered for TR2 Sports Models

FEW cars can have had so successful an introduction to competition work as the Triumph TR2 during the past season. The team prize in the Tourist Trophy can be matched by a similar award in the Alpine Trial, and individual performances of especial merit have been put up in such widely varying events as the Le Mans 24-hour race and the Production Car race which preceded the German Grand Prix on the Nurburgring.

Although shown first at Earls Court in 1952 the TR2 production was delayed for a year of intensive development which led to some considerable and valuable modifications. In production form it was first seen at the London Motor Show of 1953, and during the past year a good deal of attention has been paid to the development of optionally available accessories.

Buyers' Options

The policy of the makers has been to produce a basic sports model at the lowest possible price and to offer additional equipment to meet individual requirements. Probably the most important mechanical feature which can be selected is the Laycock-de Normanville overdrive, for with this engaged the car will exceed 107 m.p.h. and must be driven at over 80 m.p.h. before fuel consumption falls below 30 m.p.g. In addition, however, the engine can be equipped with a light-alloy sump and the chassis modified by using knock-on wire wheels, stiffer front springs, larger rear dampers, and an undershield. Available body fittings include aero screens, rear wing covers, leather upholstery, metal cockpit cover for use in competitions with driver only aboard, interior heater, radio set, hand tools, telescopic steering column, two-speed screen wipers, fitted suitcase and a modified steering wheel. This considerable variety has now been extended by a further major fitment in the shape of a glass-fibre cover which can be bolted on to turn the car into a hardtop coupé. From now on this will be available as a standard option on all TR2s, and as can be seen from the photograph it is a handsome item which blends well with the general lines of the car and at the same time provides ample rear window

BASIS of the highly successful TR2 is a light but sturdy cross-braced chassis with coil-spring independent front suspension and long underslung leaf-springs at the rear. Two inclined S.U. carburetters are used on the 2 litre o.h.v. engine.

area. It is bolted on to the main body frame and if and when it is discarded the owner can resort to the standard hood which remains in situ. The hood itself has been modified and is now fitted with quarter lights, and other small body changes are the use of quick-acting fasteners to replace the cable-operated bonnet lock, and shallower doors. The use of a sill beneath the door ensures that the latter will not foul on low kerbs.

If the car is supplied with the hard-top there come with it sidescreens which have sliding Perspex panels, and in this form nearly all the advantages of a normal coupé are attained with little additional weight and with the ability immediately to convert to competition trim if needed.

To recapitulate the main features of the design, the four-cylinder, overhead-valve 2-litre engine which develops 90 b.h.p. at 4,800 r.p.m. is joined (when overdrive is fitted) to a five-speed transmission upon which the car will do 107, 105, 79, 52, and 31 m.p.h. respectively.

An open propeller shaft drives to a conventional live rear axle with semi-elliptic springs, and although the front wheels are normally suspended by open coil springs and unequal length wishbones the hubs are linked to the latter by ball and socket joints. The frame is a box section with bracing, and the body is constructed with simple lines free from reverse curves and therefore essentially cheap to repair in the event of accidental damage. The low drag is evidenced not only by the good maximum speed on the available h.p. but also by the remarkable consumption figure of 27 m.p.g. at a steady 90 m.p.h. The combination of 34.5 m.p.g. overall with a sensibly sized fuel tank gives the car a range of over 400 miles.

The Renown model, fitted with a razor-edge four-door 4-5-seater saloon, has now been established for many years, and it is continued for a further season. The engine of this car is linked to a three-speed gearbox which, once again, can be supplemented by the Laycock-de Normanville overdrive.

The Motor Road Test No. 12/54 (Continental)

Make: Triumph   **Type:** T.R.2. Sports 2-seater (with overdrive)
Makers: The Standard Motor Co. Ltd., Coventry.

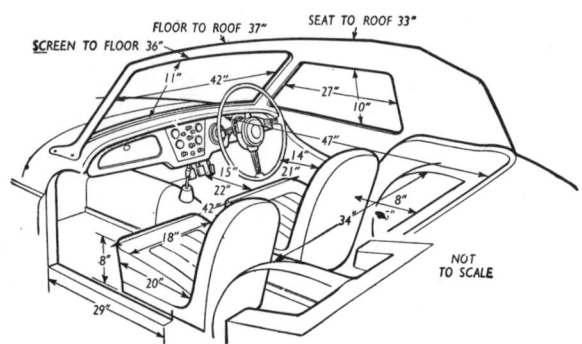

Test Data

CONDITIONS: Cold, dry weather with moderate cross wind. Belgian premium-grade pump fuel. Smooth concrete road surface (Ostend-Ghent motor road). Car tested with hood and sidescreens erect, and with tyre pressures at 28-32 lb. as advised for sustained high speeds.

INSTRUMENTS
Speedometer at 30 m.p.h. .. .. 4% fast
Speedometer at 60 m.p.h. .. .. 5% fast
Speedometer at 90 m.p.h. .. .. 6% fast
Distance recorder .. .. 1% fast

MAXIMUM SPEEDS
Flying Quarter Mile (overdrive gear)
 Mean of four opposite runs .. 107.3 m.p.h.
 Best time equals .. .. 108.4 m.p.h.
Speed in gears
 Max. speed in 4th gear .. 105.3 m.p.h.
 Max. speed in 3rd gear .. 79 m.p.h.
 Max. speed in 2nd gear .. 52 m.p.h.
 Max. speed in 1st gear .. 31 m.p.h

FUEL CONSUMPTION (in overdrive)
52.0 m.p.g. at constant 30 m.p.h.
54.0 m.p.g. at constant 40 m.p.h.
49.5 m.p.g. at constant 50 m.p.h.
43.5 m.p.g. at constant 60 m.p.h.
37.5 m.p.g. at constant 70 m.p.h.
31.0 m.p.g. at constant 80 m.p.h.
27.0 m.p.g. at constant 90 m.p.h.
Overall consumption for 1,904 miles, 55.2 gallons, = 34.5 m.p.g. Fuel tank capacity 12½ gallons.

ACCELERATION TIMES Through Gears
0-30 m.p.h. .. .. 4.0 sec.
0-40 m.p.h. .. .. 6.0 sec.
0-50 m.p.h. .. .. 8.2 sec.
0-60 m.p.h. .. .. 12.0 sec.
0-70 m.p.h. .. .. 15.8 sec.
0-80 m.p.h. .. .. 22.1 sec.
0-90 m.p.h. .. .. 30.4 sec.
Standing Quarter Mile .. .. 18.6 sec.

ACCELERATION TIMES on Three Upper Ratios

	Overdrive Top	4th	3rd
10-30 m.p.h.	—	8.6 sec.	6.0 sec.
20-40 m.p.h.	11.0 sec.	8.6 sec.	5.8 sec.
30-50 m.p.h.	11.3 sec.	8.7 sec.	6.0 sec.
40-60 m.p.h.	12.5 sec.	9.0 sec.	6.5 sec.
50-70 m.p.h.	14.2 sec.	10.1 sec.	7.0 sec.
60-80 m.p.h.	16.0 sec.	11.3 sec.	—
70-90 m.p.h.	19.3 sec.	14.2 sec.	—

WEIGHT
Unladen Kerb Weight .. .. 18¼ cwt.
Front/rear weight distribution .. 54/46
Weight laden as tested .. .. 22¼ cwt.

HILL CLIMBING in 4th gear (At steady speeds).
Max. speed on 1 in 20 .. .. 94 m.p.h. (overdrive, 82 m.p.h.)
Max. speed on 1 in 15 .. .. 89 m.p.h. (overdrive, 69 m.p.h.)
Max. speed on 1 in 10 .. .. 71 m.p.h
Max gradient on overdrive gear 1 in 11.4 (Tapley 195 lb./ton.)
Max gradient on 4th gear 1 in 8.0 (Tapley 275 lb./ton.)
Max gradient on 3rd gear 1 in 5.9 (Tapley 375 lb./ton.)

BRAKES at 30 m.p.h.
1.00 g retardation (= 30 ft. stopping distance) with 135 lb. pedal pressure.
0.97 g retardation (= 31 ft. stopping distance) with 100 lb. pedal pressure.
0.70 g retardation (= 43 ft. stopping distance) with 75 lb. pedal pressure.
0.42 g retardation (= 72 ft. stopping distance) with 50 lb. pedal pressure.
0.22 g retardation (= 137 ft. stopping distance) with 25 lb. pedal pressure.

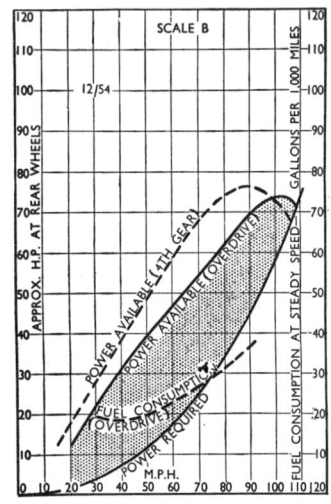

Drag at 10 m.p.h. .. .. 29 lb.
Drag at 60 m.p.h. .. .. 112 lb.
Specific fuel consumption when cruising at 80% of maximum speed (i.e. 85.8 m.p.h.) on level road, based on power delivered to rear wheels .. 0.59 pints per b.h.p/hr

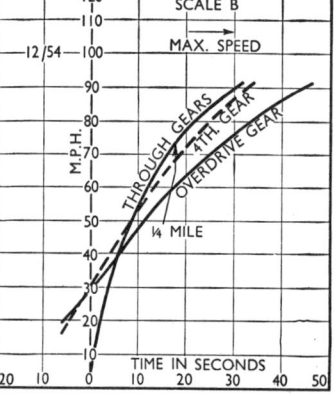

Maintenance

Sump: 11 pints, S.A.E. 30 summer, S.A.E. 20 winter. **Gearbox:** 1½ pints, S.A.E. 30 (2 pints extra on overdrive models). **Rear Axle:** 1¼ pints S.A.E. 90 hypoid gear oil. **Steering gear:** S.A.E. 90 gear oil. **Radiator:** 13 pints (2 drain taps). **Chassis lubrication:** By grease gun every 1,000 miles to 13 points and every 5,000 miles to 10 additional points. **Ignition timing:** 4° B.T.D.C. static. **Spark plug gap:** 0.032 in. **Contact breaker gap:** 0.015 in. **Valve timing** (set with 0.015 in. valve clearance). Inlet opens 15° B.T.D.C.; Exhaust closes 15° A.T.D.C. **Tappet clearances:** (Cold). Inlet 0.010 in. Exhaust 0.012 in. (for high speeds, 0.013 in., inlet and exhaust). **Front wheel toe-in:** ⅛ in. **Camber angle:** 2° positive. **Castor angle:** 1°-2° positive. **Tyre pressures** (normal use): Front 22 lb. Rear 24 lb. (increase by 6-8 lb. for sustained high speeds). **Brake fluid:** Lockheed orange. **Battery:** 12-volt, 51 amp.-hr., Lucas GTW9A/2. **Lamp bulbs:** 12 volt. **Headlamps:** 60/36 watt Lucas No. 404. **Parking, tail and number plate lamps,** 18/6 watt Lucas No. 361.

The TRIUMPH Sports 2-seater
(With Overdrive)

The Lowest-priced British 100 m.p.h. Car Displays Astonishing Fuel Economy

LOW FRONTAL AREA gives minimum air resistance but leaves useful internal roominess. High-mounted headlamps have a useful range but give a very limited sideways spread of light.

In Brief

Price (with overdrive) £635 (plus purchase tax £265 14s. 2d.) equals £900 14s. 2d.
Capacity ... ... 1,991 c.c.
Unladen kerb weight ... 18½ cwt.
Fuel consumption... ... 34.5 m.p.g.
Maximum speed ... ...107.3 m.p.h.
Maximum speed on 1 in 20 gradient... 94 m.p.h. (4th gear)
Maximum top (4th) gear gradient... ... ... 1 in 8.0
Acceleration:
10-30 m.p.h. in top (4th) 8.6 secs.
0-50 m.p.h. through gears 8.2 secs.
Gearing: 20.2 m.p.h. in top (4th) at 1,000 r.p.m. (24.6 m.p.h. in overdrive); 83.5 m.p.h. at 2,500 ft. per min. piston speed (102 m.p.h. in overdrive).

IN BRITAIN and also in many important export markets, the Triumph T.R.2 sports two-seater which we have recently tested over a distance of 2,000 miles is much the lowest-priced car which can exceed the "magic" speed of 100 m.p.h. Equipped with the manually selected overdrive fifth gear ratio which is an optional extra, it is also a car whose fuel economy verges upon the fantastic.

Engine, transmission and body designers have combined with unusual effectiveness to give this roomy two-seater car its remarkable performance, highlights of which are the ability to accelerate from rest to 60 m.p.h. in only 12 seconds, a maximum speed (with hood and sidescreens erect) of 107.3 m.p.h., and fuel consumption figures which range from 54 m.p.g. at a steady 40 m.p.h. to 27 m.p.g. at a steady 90 m.p.h. Reasonable all-up weight has been combined with very low air resistance to make this an easy-running car, and a highly efficient engine has been combined with extremely advantageous gearing to provide astonishing economy of fuel. It requires emphasis that our measured performance figures were no flash-in-the-pan, but were recorded on the way back from a visit to Switzerland without any attention being given to tappets, sparking plugs, contact breaker or other engine details for more than 1,500 miles.

Derived from the well-tried Standard Vanguard engine, the power unit of the Triumph Sports has a reduced cylinder bore which brings its size just below 2 litres, pistons giving reduced friction, a high compression cylinder head, and two S.U. carburetters. The compression ratio of 8½/1 is high enough to allow the engine to pink at low speeds on any premium-grade fuel which we tried, but neither the high compression ratio nor the modified valve timing and dual carburetters which permit higher r.p.m. to be attained have robbed the engine of its known ability to pull well at moderate speeds. An instant starter from cold, it develops enough power to give the car excellent performance when using extremely economical carburetter settings.

Delightful Gearbox

The gearbox and overdrive which have been mated to this engine are, for the keen driver, a sheer delight. The four-speed gearbox, which is controlled by a sturdy and nicely positioned central remote-control lever, has the close ratios which are expected on a sports car and synchromesh mechanism which is amply effective without being obstructive. A push-pull electrical switch engages and disengages the overdrive (which operates only when top gear is in use) positively and reasonably smoothly without use of the clutch. No automatic or semi-automatic device is provided to take over from the keen driver the pleasant duty of deciding for himself which of the five progressively-spaced gear ratios he wants to use at any particular moment. To complete the transmission, there is a clutch which is firm-acting without being so fierce as to preclude starts from rest in 2nd gear, and a hypoid rear axle which allows 100 m.p.h. to be exceeded even without use of the optional-extra overdrive ratio. This car has the pleasing characteristic of being free from "awkward speeds," of being able to produce good acceleration in one ratio or another at any pace between a standstill and 90 m.p.h. or more.

The overdrive gear fitted to the test car is an optional but extremely worth-while extra, omission of which reduces the basic price by £40 and saves a further £16 13s. 4d. on purchase tax in the case of home-market sales. Using the direct 4th ratio instead of the overdrive 5th gear only reduced the timed maximum speed from 107.3 m.p.h. to 105.3 m.p.h., but the former figure corresponds to something less than the recommended limit for sustained engine speed of 4,500 r.p.m., whereas the latter figure is attained with an engine speed of approximately 5,250 r.p.m.

CUT-AWAY DOORS have their tops padded to form comfortable elbow rests. The tonneau cover has a central zip-fastener, so that it can be used to cover one half, three-quarters or the whole of the cockpit.

FULL INSTRUMENTATION is provided, with the speedometer and "rev. counter" directly facing the driver. Other details visible in these pictures are separately adjustable seats, fly-off handbrake, central remote-control gear-lever, and push-pull overdrive switch on the right of the facia panel.

The Triumph Sports - - Contd.

Brake performance fully adequate to make the use of high speeds safe on the road has been provided on this model, which has 10-inch front and 9-inch rear brakes. The optional wire-spoke wheels would no doubt provide improved brake cooling for racing, but in normal form the car can be braked from high speed without any judder or sideways pull, and can be driven uncomplainingly down an Alpine pass in top gear. The "hanging" type pedals which go with hydraulic brake and clutch operation seem more than usually comfortable with the "straight legs" driving position of a low-built sports car, there being no inconvenient sloping toe-board to restrict heel movement. To the left of the clutch, the driver's foot rests naturally on the dip-switch. Effective in use, the fly-off pattern handbrake is so placed that it can chafe the driver's left calf.

Bodywork Details

Simple and rather unconventional in outline, the body lines of the Triumph evoke varied reactions from different people—the same may be said concerning the "geranium" paintwork of the model submitted for test. From a practical point of view, the merit of the bodywork shaping is unquestionable, there being comfortable room for two people and a fair amount of loose luggage inside the car, and a surprising amount of further luggage room in the rear locker, despite which the air resistance of the car with its hood and sidescreens erect is notably low. Like most open cars, this one loses some of its maximum speed when the hood is folded, an effective windscreen being also quite an effective airbrake, but for racing the complete windscreen can very easily be removed from the car and replaced by aero-screens.

Driven in cool spring sunshine with the hood folded but the sidescreens in position and the tonneau cover blanking off the space behind the seats, this model is certainly no more cold or draughty than most other open cars. Our test model had the optional-extra interior heater and windscreen de-mister fitted, this providing useful warmth around the legs even with the hood lowered and being able to keep the interior very snug when the hood is raised.

Two adjustable bucket seats are provided, but they fall just short of desirable comfort standards. For driving to Switzerland and back, we found that a thin rubber cushion which made up for inadequate padding over the seat springs also raised the driving position sufficiently to eliminate a "blind spot" behind the rear-view mirror, without cramping headroom below the hood for an average driver. Much care has obviously gone into details of the hood, the rubberised canvas of which removes completely from a folding frame, this unit being easily and quite reasonably quickly raised or lowered single-handed and behaving well during fast driving. A full-length tonneau cover, providing protection against rain showers or casual pilfering when the open car is parked, forms part of the normal equipment. With the hood erect, rearward vision is good, and with the car open all-round vision is of course virtually 100% clear.

At present, certain annoyances result from details of the body design, and it must be hoped that these will soon be dealt with. Extending the doors down several inches below floor level results in it being impossible to open one door when the car is parked beside the kerb in a majority of city streets. The absence of exterior door-handles, and of easy means of securing or releasing from inside the car the useful elbow-flaps in the rigid-framed sidescreens, plus the fact that a key (different from the ignition key) must always be used to open or close a capacious locker on the facia panel, are at present major irritants on a car which is in most respects perfectly suitable for year-round everyday use. A vital point of merit with this open body is that under no circumstances do exhaust fumes appear to get sucked forwards into the cockpit.

With its comfortable body, considerable luggage capacity, high performance and astonishing economy of fuel, the Triumph comes close to being a magnificent vehicle for long-distance continental travel on business or on pleasure. At this early stage in the model's life, however (the example tested had engine no. 9 in chassis no. 6), chassis qualities which are very adequate for countries such as Britain with reasonably smooth roads, proved disappointing on the fast and bumpy roads of France and Belgium.

With the 22-24 lb. tyre pressures suggested for normal use, the Triumph rides very comfortably at touring-car speeds. There is none of the traditional sports-car harshness, but there is a certain amount of roll and a good deal of tyre squeal during fast cornering. For our performance tests and most of our other driving, the Dunlop "Road Speed" tyres which are optional extra equipment had their inflation pressures increased by 6 lb./sq. in., as is

TWIN CARBURETTERS and modified valve timing are among the specification changes which allow a well-tried engine to provide sporting performance on a very low fuel consumption. Under-bonnet accessibility is commendably good.

Mechanical Specification

Engine
Cylinders	4
Bore and Stroke	83 mm. x 92 mm.
Cubic capacity	1,991 c.c.
Piston area	33.5 sq. in.
Valves	Pushrod o.h.v.
Compression ratio	8.5/1
Max. power 90 b.h.p.	at 4,800 r.p.m.
Piston speed at max. b.h.p.	2,900 ft. per min.
Carburetters	2 S.U. inclined, Type H.4
Ignition	12-volt coil
Sparking plugs	14 mm. Champion L10S (For hard driving, type L11S)
Fuel pump	AC mechanical
Oil filter	Purolator by-pass

Transmission
Clutch	Borg & Beck 9-in. s.d.p.
Overdrive (clutchless engagement)	3.03
Top gear (s/m)	3.7
3rd gear (s/m)	4.9
2nd gear (s/m)	7.4
1st gear	12.5
Propeller shaft	Hardy Spicer open
Final drive	Hypoid bevel
Top gear, m.p.h. at 1,000 r.p.m.	20.2
(Overdrive, 24.6)	
Top gear, ft./min. piston speed 33.4	(Overdrive, 40.8)

Chassis
Brakes	Lockheed hydraulic
Brake drums	Front, 10 in. x 2¼ in.
	Rear, 9 in. x 1¾ in.
Friction lining area	148 sq. in.
Suspension Front	Coil and wishbone i.f.s.
Rear	Semi-elliptic
Shock absorbers Front	Telescopic
Rear	Piston-type
Tyres	Dunlop 5.50—15
(Road Speed type on test car)	

Steering
Steering gear	Cam and lever
Turning circle: Left	32 feet
Right	30 feet
Turns of steering wheel, lock to lock	2¼

Performance factors (at laden weight as tested):
Piston area, sq. in. per ton	30.1
Brake lining area, sq. in. per ton	133
Specific displacement, litres per ton mile	2,660
	(Overdrive, 2,180)

Fully described in "The Motor" October 22, 1952

Coachwork and Equipment

Bumper height with car unladen:
Front (max.) 17½ in., (min.) 9½ in.
Rear (2 vertical bars only) (max.) 20¼ in., (min.) 11½ in.
Starting handle	Yes
Battery mounting	On scuttle
Jack	Screw type
Jacking points	On frame, reached through trap-doors in floor

Tool kit: Wheelbrace, jack, starting handle.
Exterior lights: Two headlamps, two side lamps/direction indicators, two tail lamps/direction indicators, one stop/number plate lamp.
Direction indicators	Flashing type, self-cancelling
Windscreen wipers	Two-blade electric

Instruments: Speedometer with decimal trip, tachometer, oil pressure gauge, coolant thermometer, ammeter, fuel contents gauge.
Warning lights: Dynamo charge, headlamp main beam, direction indicators.
Locks: With ignition key ... Ignition
With other key Glove locker, luggage boot
Glove lockers	One on facia, with locking lid
Map pockets	Two on doors
Parcel shelves	Nil
Ashtrays	Nil
Cigar lighters	Nil
Interior lights	Nil
Interior heater:	Optional extra, re-circulating type, with windscreen de-misters.
Car radio	Optional extra

Extras available: Overdrive, knock-on wire wheels, cast aluminium engine sump, stiffer front springs, larger rear shock absorbers, aero-screens, undershield, rear wing spats, leather upholstery, metal cockpit cover, interior heater, radio, tool roll and tools, telescopic steering column, Road Speed tyres, two-speed screen wipers, fitted suitcase, dished steering wheel.
Upholstery material	Vynide
Floor covering	Pile carpets

Exterior colours standardized (with effect from May, 1954): Signal red, pearl white, British racing green, black. Upholstery: Brown, blue or red. Hood and sidescreens: Fawn or black.

---

advised for sustained high speed driving, this also giving quicker steering response.

Over virtually the whole range of lateral accelerations used on sharp or moderately large-radius corners, this car shows a consistent but not exaggerated "understeer" characteristic, so that it is viceless right up to the limit of tyre adhesion. Only on wet and slippery roads did raised tyre inflation pressures appear to impair road-holding qualities. On really fast curves, however, it is wise to allow for the fact that due to light damping of the rear springs an unex-

LUGGAGE ACCOMMODATION inside the car is supplemented by an external locker of useful size which may be locked with either Yale or carriage keys. A separate lower compartment accommodates the spare wheel and jack.

pected bump can throw the car off its line to some extent. At speed, the steering is direct enough to transmit a fair amount of reaction to the driver's hands.

Especially with the car laden with two people and their normal touring luggage, the standard suspension seems much too lightly damped for open but badly surfaced roads, so that it can be necessary to keep the cruising speed down to 70 m.p.h. or less in circumstances when a much higher pace would otherwise be safe and economical. Unhappily, the exhaust system at present in use emits a quite ludicrous amount of noise at engine speeds around 2,400 r.p.m., and in the overdrive gear this often corresponds closely to natural cruising speeds used at night or on rough roads.

It will be noted that, amongst the items of optional equipment available for this model, stiffer front springs and larger rear shock absorbers are listed, and the latter at least are probably desirable for long-distance travel as well as for racing use. Increased damping would also no doubt minimize unexpectedly vigorous "shake" of the front end and scuttle at speeds around 75-80 m.p.h. which became increasingly evident towards the end of our extended Continental test, although the impression is formed that further stiffening of the frame or scuttle may be desirable.

Apart from chassis imperfections on this early example, the Triumph has merit for fast business travel as well as for sporting use. Almost completely weatherproof, even if slightly undignified to enter when the hood is raised, its astonishing economy at fast cruising speeds is backed up by the provision of a large fuel tank which enabled us to drive quite rapidly from Geneva to Calais, over 450 miles away, without any re-fuelling stop. Secured by two locks instead of the usual one, the alligator bonnet gives very good access to the power unit. The light and high-geared steering provides an extremely compact turning circle. A full set of instruments has been sensibly arranged on the plastic-covered facia panel.

Although we have felt obliged to criticize some details and characteristics of the Triumph T.R.2 Sports two-seater in quite emphatic terms, we nevertheless rate this as not merely the best sports car available at its price, but also as one of the most promising new models which has been introduced in recent years. Not pared down to minimum weight especially with a view to use as a competition car, this model offers a combination of comfort, economy, speed and sheer enjoyment of travel in a responsive open two-seater, which should assure it of very large sales in many parts of the world.

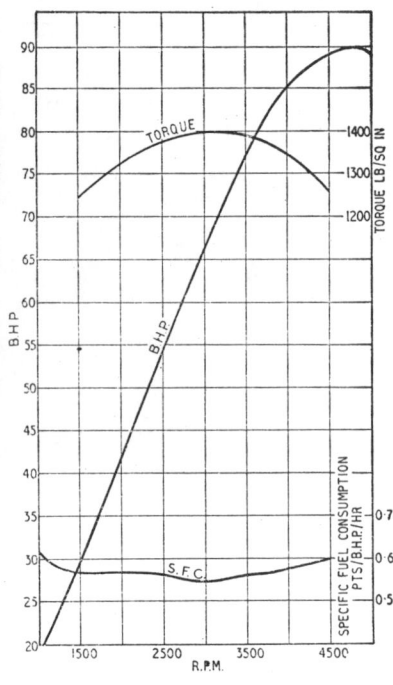

A painted and polished show chassis which illustrates the compact layout of the Triumph TR2. The engine is placed low down and well forward in the frame, and the cooling fan runs in front of the centre section of the track rod and the tubular tie bar placed between the front spring abutment brackets

A TRIUMPH OF DEVELOPMENT

THE STORY OF THE TR2   By JOHN RABSON, A.M.I.Mech.E.

WHEN a large manufacturer such as the Standard Motor company decides to increase its range by the addition of a model produced in small quantities (in comparison with the Vanguard, for example) and at a low price, there is a limit to the amount that can be spent on tools and equipment. It would not have been economical, therefore, to start with a clean sheet of paper and design a sports car that was new from bumper to bumper, develop it, and finally put it into production. Instead, the car had to be based on a number of existing items. This governed the whole layout of the TR2, and at the same time enabled it to be produced competitively.

Early in 1952 it was decided that a cheap 90 m.p.h. sports car should be produced, the major components being basically similar to those of existing models: the Vanguard engine, the immediate post-war Standard Eight frame, and the Mayflower front suspension and rear axle. "Basically similar" is not intended to mean identical, yet if a large degree of general similarity is maintained, it is possible to machine components with the aid of existing jigs and fixtures, although the actual strength of the components may vary considerably.

There were no existing units that could be used for the body, so this had to be completely new. However, the choice of style was severely restricted by the money that could be spent on press tools. This basic specification influenced the whole character of the car, and explains the course of the development programme.

In car development one thing is closely related to another; for example, if the power output of the engine is increased it is usually necessary to improve the brakes. Indirectly, almost all the development work carried out on the TR2 has been a result of improved performance brought about by increased power output.

First of all, therefore, came the development that converted a Standard Vanguard engine developing 68 b.h.p. at 4,200 r.p.m. in "bare" engine condition or 65 b.h.p. in "full road trim," to the Triumph TR2 power unit which, in its latest production form, develops 90 b.h.p. at 4,800 r.p.m.

The story becomes even more interesting when it is realized that the original target figure of 90 m.p.h. was very considerably exceeded, with the result that the TR2 is most certainly one of the cheapest production cars with a genuine three-figure maximum speed attainable in touring trim.

To enable the Triumph to be used in competition in the up to 2-litre class, it was necessary to reduce the engine capacity from the 2,088 c.c. of the Standard Vanguard engine, and a figure of 1,991 c.c. was obtained by reducing the cylinder bore diameter from 85 to 83mm, while retaining the 92mm stroke. This was a simple matter on a Vanguard type engine, as it was necessary only to fit new liners with a smaller bore instead of redesigning the cylinder block, which might be necessary on an engine with integral bores if the cylinder walls were to be prevented from becoming too thick. At the same time the cylinder head was modified to give a compression ratio of 7.5 to 1. New manifolds were fitted, the single downdraught Solex carburettor being replaced by twin S.U. instruments (H4) having 1¼in diameter bores. On the brake, and with manual ignition control, this resulted in a power output of 73 b.h.p., an increase of 8 b.h.p.

The manual ignition control should perhaps be explained:

in engine development work it is usual to run initially with a manual ignition setting to obtain the correct distributor advance curve; when development has been finalized, the necessary automatic advance curve characteristics are produced from data obtained on the test bed.

There was still a long way to go before the 90 b.h.p. of the production target was obtained. The next stage was further to increase the compression ratio to 8.6 to 1 (slightly higher than the final production figure), which produced 79.7 b.h.p. After this, work on the engine was concerned primarily with camshaft and valves. First the valve lift was increased from 0.36in to 0.375in, still retaining the familiar 10-10, 50-50 valve timing. Then, keeping the 0.375in valve lift, the timing was modified to give increased overlap, with the inlet valve opening 15 deg before top dead centre, and closing 55 deg after bottom dead centre, and the exhaust valve opening 55 deg before bottom dead centre and closing 15 deg after top dead centre. This resulted in an output of 84.1 b.h.p., another useful step in the upward march of power.

The third stage was to increase the diameter of the inlet valve from 1½ to 1 $\frac{11}{16}$ in diameter, and suitably to enlarge the inlet ports. This modification resulted in another 3 b.h.p., bringing the total up to 87, while further development work and final tuning on the carburettor needles and distributor curve resulted in an output of around 90 b.h.p., the present production figure. The story is not quite as simple as it sounds; during the development stages a number of difficult problems had to be solved to ensure the complete reliability of the unit, for both normal driving and severe competition work, before it reached the hands of the owner.

In designing any production car—saloon or sports car—weight and cost are kept to a minimum; consequently components that give complete reliability in an engine developing 65 b.h.p. will not necessarily do so if the power output is increased to 90. The main function of vehicle development is to find weaknesses by very strenuous testing and

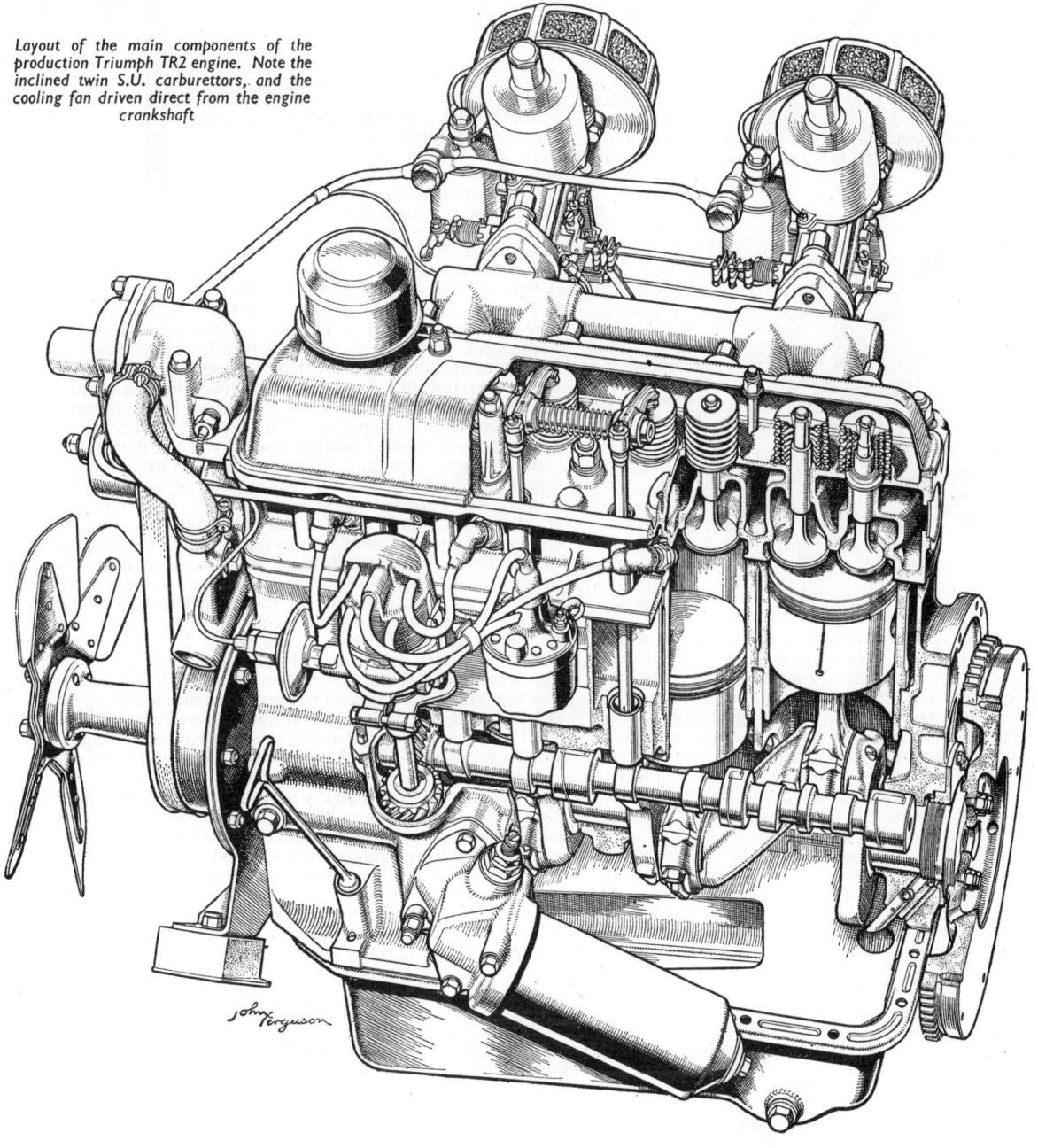

Layout of the main components of the production Triumph TR2 engine. Note the inclined twin S.U. carburettors, and the cooling fan driven direct from the engine crankshaft

A TRIUMPH OF DEVELOPMENT . . . . . continued

make the necessary modifications. At the same time, the general stressing in the major engine components must be reviewed, and any redesign dictated by increased loading must be incorporated, to ensure that no component is overstressed.

Besides ensuring that the power unit operates satisfactorily on the test bed, it is necessary to make certain that it will do so equally well in a car on the road. The difference between the two sets of operating conditions can vary tremendously. Although it is possible to vary both the air temperature and conditions of loading when an engine is on the test bed, the power unit in road work is subjected to changes of air velocity which further complicate the problem, as does the flow around the engine compartment.

The raising of power output and b.m.e.p. increases the gas pressure acting on the pistons, and the TR2 pistons have very stiff crowns. The improved power output also increases the force tending to lift the cylinder head from the block—a force that must be resisted by the cylinder head studs, or the head joint gasket will have a very limited life.

The first weak link in the course of testing was found to be the cylinder block; cracking occurred below some of the intermediate cylinder head stud bosses. On the original engine, lugs were provided locally towards the top of the cylinder block, and the trouble was remedied by altering

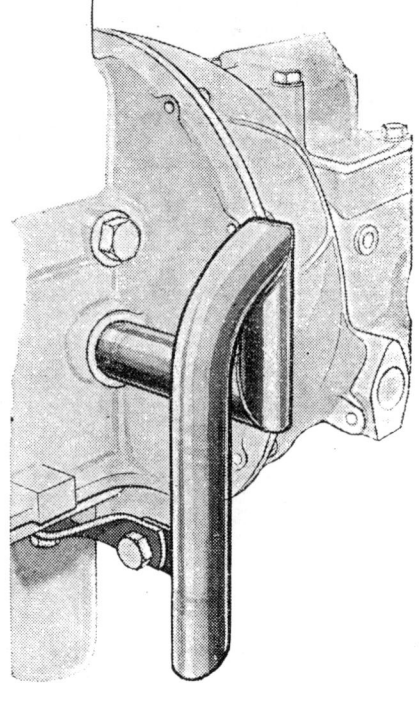

Much development work was done on the crankcase breather to prevent oil being flung out because of rapid cornering, and also to prevent it from being drawn out by air passing over the end of the tube when the car was travelling at maximum speed

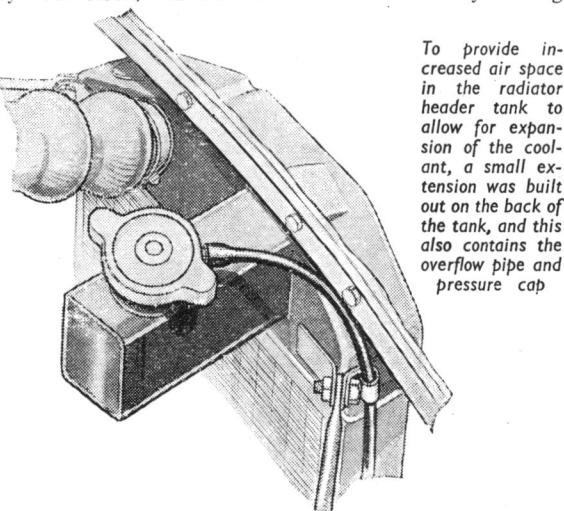

To provide increased air space in the radiator header tank to allow for expansion of the coolant, a small extension was built out on the back of the tank, and this also contains the overflow pipe and pressure cap

the main block casting (as shown in an illustration) so that the lugs were extended right down the inside of the water jacket wall to the very stiff crankchamber below. With this arrangement the studs are actually screwed into the upper part of the crankchamber, so that tightening down the head places the water jacket wall in compression, as opposed to tension with the previous arrangement. This modification proved to be entirely satisfactory. But although it solved the cylinder block problem, it did not remedy cylinder head gasket trouble completely.

The bottom joint of each pair of cylinder liners is sealed by a thin figure-of-eight gasket; on the Vanguard this was made of wire-reinforced jointing material, and the thickness of the bottom joint washer must be such that the cylinder liners stand slightly proud of the top joint face, ensuring that the top gasket is completely nipped around the combustion chamber when the head is tightened down.

Because of the increased stud loading imposed by the improved power output, there was a tendency for the bottom gaskets to collapse, and this, in turn, caused the cylinder head gasket to blow. Resin-coated steel figure-of-eight gaskets were substituted to form the bottom seal, and these completely cured the trouble. However, it was necessary also to reduce production tolerances on the various components concerned, and also to maintain the cleanliness that is vital during assembly.

The increase in power output was, of course, accompanied by an increase in engine speed, and to prove the power unit, prototype cars were driven for many hours at maximum speed around the banked high-speed circuit of the Motor Industry Research Association proving ground at Lindley. Initially, between two and three hours' running at a sustained speed equal to 5,200 r.p.m. in direct top gear brought about big-end bearing failure, and the first modification was to substitute indium coated lead bronze bearings for normal white metal bearings on the big ends; no trouble was experienced with overdrive. Although this effected some improvement, it did not completely cure the big-end trouble, and attention was next given to the lubrication system. First, modifications were made to the *main* bearing shells with the idea of permitting oil to pass more quickly from the crankcase via the main bearings to the crankshaft, and from there to the big-end bearings via crankshaft drillings in the normal way. This was not a complete answer, and it was not until the *crankshaft* drillings were modified that the problem was completely solved. The crankshaft was originally drilled through from the main bearing to the big-ends. This resulted in considerable loss of oil owing to centrifugal action from the higher crankshaft speeds combined with the increased bearing clearances necessary with the lead-bronze bearings.

If the crankshaft drillings are modified so that the oil is discharged at a point around the periphery of the big-end bearing that is closer to the centre line of the crankshaft, this effect will be considerably reduced, and this system of cross-drilling has been employed on the Triumph. The outer end of the main oilway, where it breaks through to the surface of the big end, has been plugged, and a cross-drilling from the main oilway added. Further to improve the spread of oil around the bearing, the edges of the holes

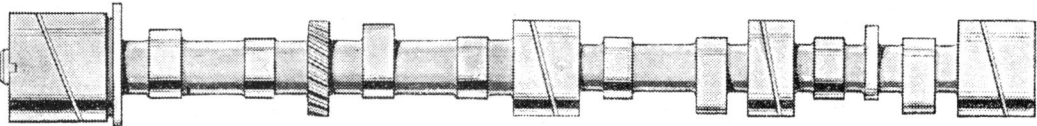

To overcome valve trouble caused by camshaft deflection, the front half of the shaft was increased in diameter. Increasing the diameter throughout the shaft did not cure the trouble

The four cylinder-head stud bosses placed between each pair of bores have been extended to the bottom of the block, so that the studs are, in fact, screwed into the top of the crankchamber

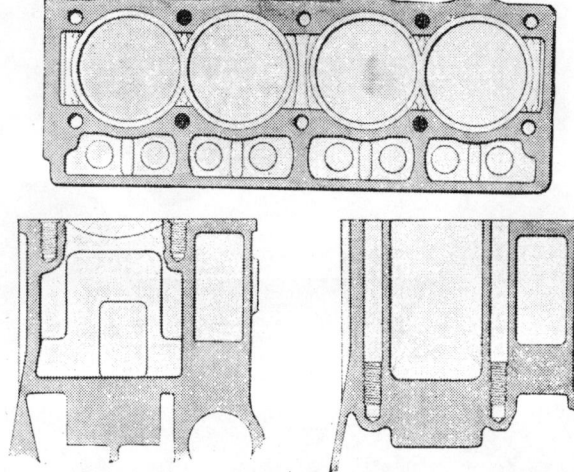

have been "shelled" and this process also prevents "cutting up" of the bearing shell surface by foreign matter.

It was mentioned earlier that, with the increased power output, it was necessary to re-stress some of the major components, and the connecting rods came into this category. Although the rods did not fail, it was thought desirable to increase their stiffness in order to provide a greater factor of safety; at the same time the system of location between rod and cap was improved. Both the big end and cap were stiffened considerably to reduce the measure of flexibility that had caused fretting between the backs of the big-end bearing shells and the rods. The diameter of the big end bolts was increased to 7/16in, and the rod and cap location was provided by a single tubular dowel in place of a double location, a method that ensures greater accuracy and assists production. Further, the tubular dowel prevents shear loading on the bolt.

On the Road

After each modification was proved on the test bed it was tried out under high speed road test conditions, the car being driven very much harder than would be possible on normal roads in this country. These tests brought to light additional faults, including troubles with engine breathing, crankshaft oil seals, and valve gear.

Perhaps the most important, and certainly the most interesting, problem was that connected with valve gear, as it underlines the point that, in all development work, it is necessary to analyse carefully what is taking place and how it is causing difficulty, before the trouble can be remedied effectively.

At this stage of development, when valve gear endurance tests were being conducted, number 1 exhaust valve failed after 246 miles of running at over 5,000 r.p.m. If repeated tests of a similar nature had produced failure of the exhaust valve in other cylinders, it would seem logical to assume that the exhaust valves were not strong enough for the task which they had to perform. The remedy would be to fit stronger valves.

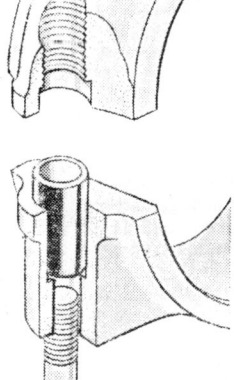

The two halves of the connecting rod are located by a single tubular dowel; this also reduces shear loading on the bolt

The fact that failure was *not* general meant that a thorough investigation of the operation of the valve gear was required, and this was carried out on a special valve test rig enabling the effects of valve gear under high speed operation to be observed by stroboscopic and electronic test equipment. These indicated that at speeds around 6,000 r.p.m. the camshaft was bending 0.019in on number 1 exhaust cam, but only 0.008in on number 3 exhaust cam. It should be emphasized that no trouble of this kind was experienced on the normal Vanguard engine which, of course, does not operate at nearly such a high speed. One remedy that was tried was to increase the shaft by 1/8in in diameter throughout its length, but this did not solve the problem, and no reduction in bending was obtained. The final solution was to increase the diameter of the front half of the shaft, leaving the rear section the original size; at the same time the valves were strengthened as an added precaution.

For continuous operation at very high speed, it was found necessary to improve both the oil sealing and the crankcase breather arrangements. At the rear end of the crankshaft the normal arrangement of a plain surface in conjunction with an outer return oil scroll was modified so that a scroll was provided on the shaft also. This effectively prevented high speed oil leaks from the rear end of the crankshaft; but there was still loss of oil via the crankcase breather. As engine speed and performance are increased the problems connected with crankcase breathing become more complex; the chamber must be effectively ventilated to prevent the build-up of pressure, yet the increase in road speed may cause the air current over the outer end of the breather pipe to have an extractor effect. An illustration shows the final arrangement, which provides an unobstructed path for the gases; the inclusion of a section of vertical pipe away from the crankchamber tends to prevent oil from being flung out, and also helps to condense oil vapour and enable it to run back into the sump.

Because of the low bonnet height, the radiator is placed low down in front of the engine, with the result that the normal arrangement of belt-driven fan could not be used, so the fan was placed on the front end of the crankshaft. Although this proved to be satisfactory as regards radiator cooling, torsional vibration of the crankshaft caused breakage of the fan, and to cure this, rubber bushes were inserted to provide some insulation between the shaft and the fan.

Following these modifications the power unit proved to be completely reliable, and the car could be seen on the banked circuit at M.I.R.A. consistently lapping the circuit at 100 m.p.h. for hours on end, the only stops being for refuelling every three hours. However, it was also necessary to make a considerable number of modifications to the chassis to keep pace with the ever-increasing performance. Details of these will be explained in Part II.

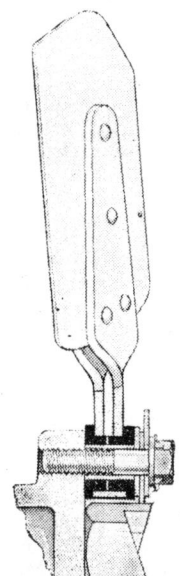

The fan is mounted on the front end of the crankshaft, and to prevent fan blade fatigue caused by vibration, rubber bushes are used in the mounting

A TRIUMPH OF DEVELOPMENT

THE STORY OF THE TR2— PART II (Conclusion)

By JOHN RABSON, A.M.I.Mech.E.

How the various stages of development have changed the appearance of the TR2. In addition to the hard top, the third picture shows the latest arrangement of the doors, which now have fixed sills. This car also has centre-lock wire wheels, available as an extra

IN Part I it was explained how the Triumph TR2 was developed, and its power output increased from 65 to 90 b.h.p. It was, of course, necessary to develop and modify chassis components in step with the ever-increasing "power available." With the Triumph, perhaps even more development work than usual was required because of the very considerable amount of extra power; whereas the original intent was to produce a 90 m.p.h. vehicle, it was soon found that it was a 100-mile-an-hour-plus two-seater in standard trim, while 124.095 m.p.h. was reached over a measured mile at Jabbeke in speed trim.

Although the original brakes would have been satisfactory on the car with its original power output, as the engine performance increased it was found that they were inadequate; fade was experienced, and pedal pressures were too high. The car had 9in drums with an effective lining width of 1¾in at both front and rear; the shoes were hydraulically operated, with two leading at the front, and leading and trailing shoes at the rear. After lining tests, the next modification was to increase diameter and width of the front drums to 10in × 2¼in, and, compared with the 9in front brakes, there was a marked improvement, particularly with the use of Mintex M20 linings in the front drums. A new problem—that of brake judder—was overcome by improving the inner surface finish of the drums. By now the brakes were more than adequate for normal fast touring, and in fact no fade was experienced during *The Autocar* Road Test. To give an added safety margin for the strenuous competition work to which the cars might be subjected, the rear brake drum size was increased at a later date to 10in × 2¼in.

An example of the utilization of existing tools and equipment to produce new components is shown by the transmission. The Vanguard has a three-speed gear box with synchromesh on all forward ratios; inside a casing of similar size a new gear box was designed for the TR2 having four forward speeds with synchromesh on top, third and second gears. In place of the steering column change mechanism, a neat, central remote control was produced which is positive and particularly pleasant to operate.

An interesting feature of transmission was the extensive use made of the Laycock-de Normanville overdrive in the development stages, as it permitted high-speed endurance tests to be carried out at engine speeds lower than would be possible in direct top gear with the same axle ratio. Before the engine had reached its final stages of development it was found possible, for example, to run for a three-hour period, averaging 100 miles an hour in overdrive, whereas a similar test in direct top gear would have produced engine failure (as explained in detail in Part I).

The final drive unit first used was similar to that of the Triumph Mayflower. Much development work was done on this component both on test rigs and on a complete car. As a result of 100-hour rig tests at 5,000 r.p.m. (engine speed), the crown wheel bolts were increased from 5/16in to $\frac{3}{8}$in diameter, and these were secured by plain washers and lock tabs. Following high-speed cornering during road testing, it was found that the centrifugal force caused the oil to be thrown along the axle casing tubes and into the wheel bearings, diluting the grease, which was finally washed out. This was rectified by fitting traps to prevent the oil and grease from coming into contact.

To increase their life, the number of taper rollers in the differential bearings was increased from 16 to 18; much work was done to determine the correct fit between the bearings and their housings, and the effect of pre-load was also investigated. It was found that with both the 16 and 18 roller bearings, if the cup was an interference fit in its housing, slight pitting occurred after 50 hours' test, and after 100 hours the track of the cup failed through fatigue; however, if the clearance were increased so that the cup was a slack fit inside its housing, no failure was experienced with the same pre-load.

The story of the frame is in many ways similar to that of the brakes—the original design utilized chassis frame members similar to those used on the immediate post-war Standard Eight, suitably modified at the front for the attachment of wishbone and coil spring front suspension. This

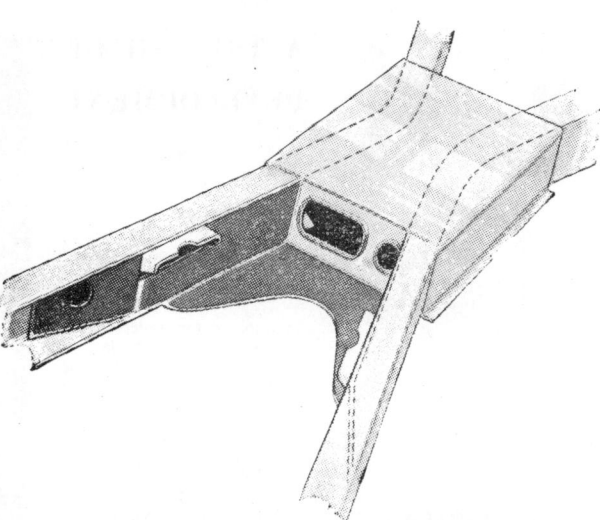

The centre of the frame was stiffened by a welded-on gusset plate extending forward from the centre section of the cruciform, while additional stiffening plates enclosed the corresponding portion of the channel section cruciform members to form box sections

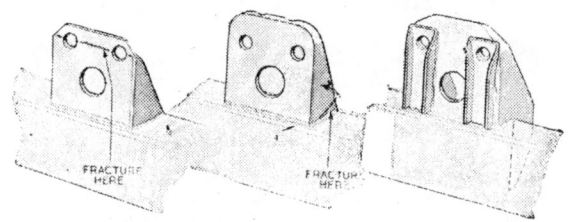

The rear damper mounting brackets required considerable stiffening as the performance of the car increased, and this illustration shows the three stages of development

resulted in a compact and lightweight frame that could be produced cheaply, and sufficiently strong for the car as it was originally designed.

When it became evident that the performance would considerably exceed the original estimate, it was found necessary to re-design the frame completely; to facilitate production and economize in jigs and tools, this was achieved without altering the major dimensions. It now had box section side members, and extra stiffening was provided at the centre of the cruciform. The rear end of the chassis is, of course, underslung, and consequently the propeller shaft runs above the centre section of the frame.

A considerable improvement in rigidity was achieved, but the company were still not yet satisfied that this basic structure was strong enough to withstand rough service in countries such as Belgium, where there is still a good deal of rough *pavé*. More testing in Belgium showed the need still further to stiffen the cruciform by welding on an extra gusset plate at the front, and enclosing the front cruciform members locally to form a box section. The rear damper bracket, too, was modified to prevent failure at the attachment point.

Attention could now be turned to the suspension units, and again, because of the considerable increase in performance, the pressed steel lower wishbones used on the Triumph Mayflower were replaced on the TR2 by forged components. The inner rubber suspension pivot bearing bushes of the lower wishbones were replaced by nylon bushes; this resulted in greater accuracy of wheel movement by obviating unwanted flexibility, and at the same time increased the bearing life. The front stub axle flanges were also made stiffer, although no actual failures were experienced during the development stages.

The car was first seen by the public at the London Show in 1952; at this time the power unit was developing 75 b.h.p. The body, with the exception of the front end styling, differed from that used on the production version which was shown at Earls Court the following year.

The present production body is the logical outcome of reducing cost to the minimum. Production was originally laid out for relatively small quantities, consequently tool costs had to be kept down, while on any vehicle in the sports car class it is necessary to provide the highest power-to-weight ratio, so the body must be made as light as possible.

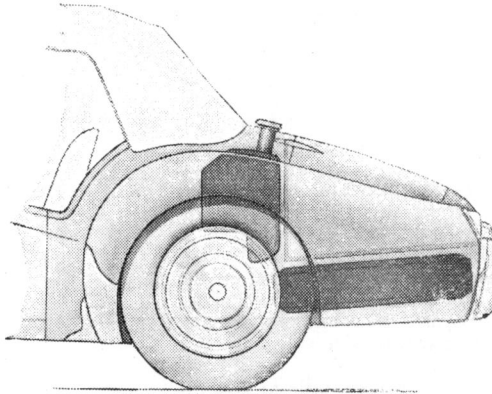

The first major modification to the body consisted of restyling the rear end so that the fuel tank was mounted over the top of the rear axle, and an external luggage compartment provided on top of the spare wheel locker

A TRIUMPH OF DEVELOPMENT
... continued

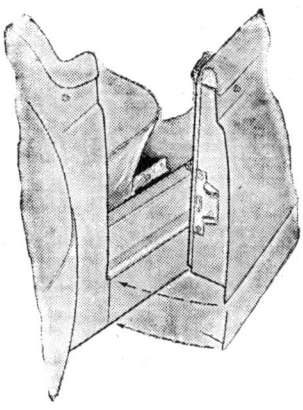

To improve the general body stiffness and provide increased kerb clearance, the doors were modified and a fixed lower sill incorporated

This high angle shot of the TR2 in Jabbeke trim shows how the cockpit was almost completely enclosed, and a small Perspex deflector fitted in place of the normal screen. Both front and rear bumpers were removed

Therefore the body was designed so that no double-action presswork was required, and all the panels were joined on the centre line—the front wing panels, for example, have a joint line on the top of the wing. Tool considerations also dictated the arrangement of the head lamps in the front panel. To keep weight down, the front and rear overhang was kept as short as possible, while to reduce body width cutaway doors were adopted.

As a result of consumer reaction at the '52 Show, a number of modifications were made to the body, the most important being the rearrangement of the rear components to enclose the spare wheel and provide a separate luggage compartment; the body width was increased slightly to give

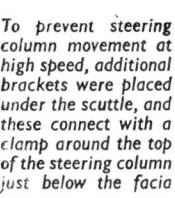

To prevent steering column movement at high speed, additional brackets were placed under the scuttle, and these connect with a clamp around the top of the steering column just below the facia

more elbow-room. In its modified form the body was subjected to many thousands of miles of endurance testing as the power output of the engine was steadily raised, and it was found necessary to make a few alterations.

Looking at the windscreen of the current TR2, a casual observer would conclude that the glass was flat and, in fact, flat glass was used on the prototype cars, but when the vehicle was lapping at speeds of 100 miles an hour and more it was found that wind pressure on the screen was sufficient to cause bowing. The screen was given $\frac{3}{4}$in camber, so that wind pressure would tend to straighten the glass and press it more firmly into the screen frame.

It was also necessary to stiffen up the hood frame and its attachment points, particularly where the front of the hood is attached to the top of the screen; the deflection caused by the wind had produced a noticeable gap between the hood and the top of the screen, so the frame and toggle attachments were stiffened. To prevent air leaks at high speed, brought about by sidescreen deflection, the leading edge of the screen was fitted into a channel section on the sides of the screen frame. This was arranged so that, as the door was closed, the two parts came into engagement, with the result that the sealing was considerably improved and screen deflection eliminated.

Slight modifications were made to the scuttle to give increased clearance for the gear lever and passengers' legs, and an extra stiffening bracket was provided to prevent steering column vibration. It was in this form that the body finally went into production, and since production commenced very few modifications have been made, the only significant one being to the doors. On the early production cars the doors did, in effect, cut the body panels into three parts; later these were modified so that there was a substantial sill joining the front and rear portions of the body. In addition to improving the overall stiffness, this modification gave much improved clearance, so that the respective door could be opened when the car was parked close to a high kerb.

An optional extra in the form of a glass fibre detachable hard top was also made available, so that the car could be converted into a fixed head coupé for the winter months.

No account of the TR2 development would be complete

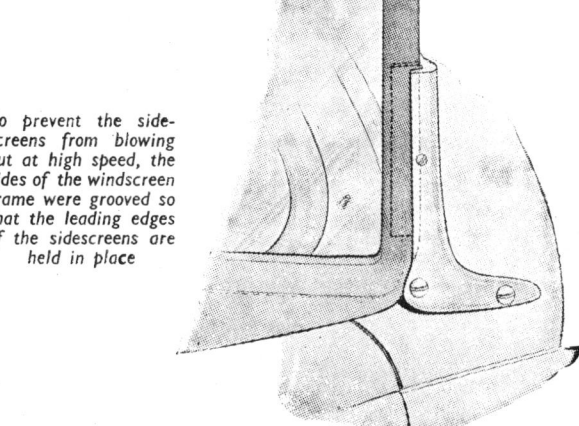

To prevent the sidescreens from blowing out at high speed, the sides of the windscreen frame were grooved so that the leading edges of the sidescreens are held in place

A TRIUMPH OF DEVELOPMENT .... continued

without mention of the development work that was necessary to prepare the car for competition, and for record-breaking runs such as the test that took place on the Jabbeke motor road in Belgium in 1953. It may be thought that, for this run, the engine had been specially tuned, but this was not so, for the regulations required the engine to be dismantled after the test for inspection by the scrutineers. How was the extra speed obtained? The runs made at around 124 m.p.h. were with the car in speed trim; this consisted of removing the normal hood and sidescreens and the front and rear bumpers, and fitting a complete cockpit cover and an undershield. The effect of these modifications was to reduce the frontal area considerably by removal of the

Degrees of saving in frontal area achieved by discarding the standard windscreen can be seen in this diagram. The Perspex cowl was used, in conjunction with a lowered driving position, for the high-speed runs at Jabbeke

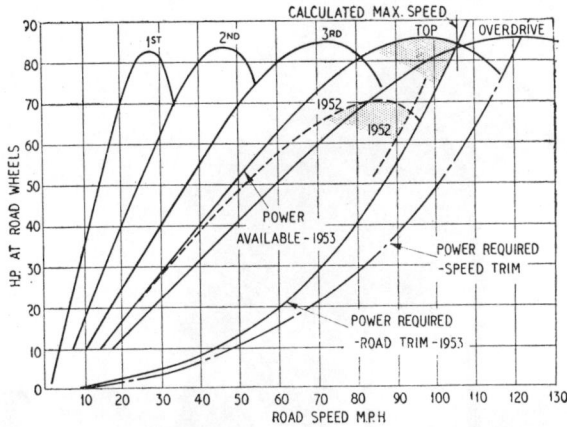

This diagram shows the power available and power required at various engine outputs and with different equipment. Note that in standard touring trim with the hood up the use of overdrive does not increase the maximum speed, yet in speed trim there is a substantial gain in maximum speed when overdrive is used.

windscreen and cut down drag by fitting an undershield and cockpit cover.

The increase in speed brought about by removing the windscreen was demonstrated at Jabbeke when subsequent runs were made with the normal windscreen and hood and sidescreens in position. In this condition—in standard touring trim, but with the addition of an undershield—a mean speed of 114.213 m.p.h. was obtained over a measured mile. By comparison with figures obtained in *The Autocar* Road Tests, it will be seen that the extra equipment considerably improves performance under a specific, limited set of conditions; for very short, high-speed runs the use of an undershield would be advantageous.

Disadvantages

The effect of the undershield under competition conditions over a long period on the temperature of the major chassis components was investigated; it was found that undershields should not be used except for short duration, record-breaking runs. By carrying out tests of this type, and passing the knowledge on to competition-minded owners of their cars, the company not only ensures the reliability of its products but also saves the private owner much expenditure.

The Triumph TR2 has established itself very quickly as a thoroughbred sports car, capable of competing with cars selling at a very much higher price, and it would be hard to find a sports car that provides better value for money. As a result of experience gained during its development stages, the company has collected much information, which can also be incorporated in the series production saloons such as the Vanguard; in fact, modifications have already been made to the Vanguard as a result of TR2 engine development, thereby increasing efficiency and reliability.

Using a casing dimensionally similar to that of the Vanguard three-speed gear box, for the Triumph a four-speed unit was developed having a neat remote control gear change

FROM ONE GENERA...

A Continental Assignment Shows the
Suitable for Fast, Long-Distance Touring

"... 100 m.p.h. was indicated ... and 64 miles were covered in

THERE were good reasons for finding an alternative to my Triumph 2000 Roadster, for reporting on the southern section of this year's Tulip Rally. The car is six years and 72½ thousand miles old. Much of the mileage has been along the stony ways of reliability trials; thousands more have been spent as the tug part of a caravan outfit, which tends to age a vehicle; for the remainder, because the exigiencies of journalism usually demand that one shall endeavour to arrive at point B from point A yesterday, it has been driven more hard than soft. Had we been able to do our journey at leisure, there would have been no hesitation in using the 2,000, but to average something over 300 miles per day for five consecutive days with the absolute necessity of no involuntary halts and the need to drive to a fairly tight schedule was, one felt, asking rather a lot of the old lady.

Then there was the question of thirst. Cruising at 60 m.p.h. on normal roads, the Roadster, with its old-type Vanguard engine, consumes, on average, a gallon of petrol for each 19 miles, and nothing I have been able to do will materially better that figure. In mountains, where much of our motoring lay, the consumption would be correspondingly greater. Therefore, if only considered on the grounds of estimated fuel consumption into limited business car allowance, it was "no go."

An offer by the Standard Motor Co., of a representative of the younger generation in the form of a TR2 with overdrive, in many respects seemed a reasonable solution. When tested by *The Motor* a similar car showed an overall fuel consumption of 34.5 m.p.g. and 43.5 at steady 60 m.p.h. At Le Mans last year, a privately-entered example finished 15th, covering 1,793 miles in the 24 hours and averaging 34.68 m.p.g. for the total distance. Thus, from economic, mileage-covering and reliability points of view, the problem was as good as solved. There remained the question of luggage stowage. Although on a job such as ours one does not travel with an extensive wardrobe, there is a limit to paring down, particularly in view of the variation of weather conditions generally experienced on "The Tulip." In addition, the co-driver also being photographer, room had to be found for all the paraphernalia without which it seems even a miniature camera cannot be operated efficiently. But, apart from protective clothing in the way of duffel coats and waterproofs, for which there was plenty of room behind the seats, the whole lot went under lock and key in this small sports car's luggage boot.

Up to then, co-driver's motoring had been almost solely in cars of much more plebeian characteristics, so it seemed a good idea to let him drive the Triumph to Dover to get the feel of it. As his dimensions run rather to girth than height, the normal seating position was a bit too low for comfort; this was soon put right with the aid of a domestic cushion. Then off we went.

Conversion Course

It takes a few miles to get through the transition stage from a family saloon weighing 21¼ cwt. propelled by 50 b.h.p., which is what co-driver usually uses, to 17¾ cwt. with 90 b.h.p. on tap. Quite soon, however, frown of concentration gave way to happy smile, a snatch of song burst forth and, metaphorically, his cap went on back to front.

Some necessary carburetter adjustment, indicated by very erratic idling during a long hold-up which sent engine temperature considerably above normal, was easily made, the twin SU's being very getatable through the

...TION TO ANOTHER

...umph TR2 to be as
for Competition Work

By E. H. ROW

...ce of an hour without consciously trying."

"... the whole lot went under lock and key in this small sports car's luggage boot."

liminary rally routes, with a stop *en route* at Rheims to fix accommodation at the Hotel Welcome for our return journey.

With the TR2's capabilities, it seemed pointless to potter painfully over the cobbled abomination of N43 when the better-surfaced way through Abbeville, Amiens, Montdidier, Compiègne and Soissons offered so much more in the way of pleasant motoring. The sun shone, the Triumph cruised happily down the poplar-lined roads with the speedometer needle steady on the "80" mark and the tachometer showing 3,250 r.p.m. in overdrive top. It was probably these conditions, coupled with the soporific effect of a good lunch, which led to the discovery that, if you aren't too tall, a comfortable and almost fully recumbent position can be achieved in the TR2 with the passenger's seat back to the fullest extent.

big bonnet opening. From then on, to the end of the trip, no further recourse to the tool roll was necessary.

Customs officials are as unpredictable as the weather. For years co-driver (hereinafter to be known as C.D.) has been crossing the Channel on his lawful occasions with the same camera and bits and pieces in the same case and with the same form to tell whosoever it may concern all about it. This time, however, it appeared that the form wasn't the right Form at all: that an entirely different FORM was what they really needed and, while they'd let it go this time, he'd better see that he got the proper one before again leaving British shores with the tools of his trade. Thus, although we were quite early on dockside parade, instead of making a smart getaway from Boulogne, through the "Lord Warden's" arrangement of "first on, first off," we were among the last to leave the boat: which shows the importance of knowing the form.

For that night our destination was Nancy, an intermediate control on the London, Paris and Berne pre-

"... flat-topped trees and the Grand Hotel G. Ripotot ...", outside which the TR2 makes an interesting contrast with one of the fixed-top models produced in the Belgian Imperia works on the Triumph chassis.

June 22, 1955

From One Generation to Another - Contd.

Impressed by previous travellers' tales of what can happen by disobeying the now fairly stringent French *Code de la Route*, we were extremely careful not to transgress overmuch. None the less, our average speed for the 180 miles between Montreuil and Rheims was something in excess of 50 m.p.h., and along one favourable stretch over 100 m.p.h. was indicated on the unchecked speedometer and 64 miles were covered in the space of an hour without consciously trying.

There was still plenty of evening left when, with 334 miles behind since just before midday, we pulled up outside the Grand Hotel in Nancy's Place Stanislas, bounded by magnificent eighteenth-century buildings, illuminated fountains and brightly lit cafés full of music and people.

Easy to Handle

That we were pleasantly tired is not surprising; there was, however, none of the "Thank goodness to get out of that thing" about it, which may surprise those who regard a sports car as a device only suitable for competitive motoring by the reasonably tough. Actually, although the TR2 suspension is firm enough for all fast motoring needs, the car rides steadily and the individual bucket seats are so planned that one is comfortably held and supported. Nor is the physical act of driving at all wearisome. The steering is light, decently high geared and has just the right degree of under-steer: gear changes like a knife through butter are easy with the short, centrally located lever. But the real joy, particularly on French roads, is what C.D. insisted on calling the "electronic gear shift"—the switch-operated engagement and disengagement of the Laycock-de Normanville overdrive top. Unlike on earlier models, the control on the present TR2 is set nice and handy on the right-hand side of the facia panel where it can be worked with a couple of fingers without taking one's hand off the wheel. Cruising down the straights at 80-85, it was "flick" down to normal top for fast bends, accelerate round and "flick" back again into overdrive; just as simple as that—no clutch, no throttle work.

The economy resulting from an overdrive, which at 60 m.p.h. reduces r.p.m. by approximately 500, compared with normal top, is, of course, of tremendous help, with French petrol the price it is, in overcoming the handicap of a limited car allowance. Looking back at my diary of the journey, I see that, despite our fair daily mileage, there was seldom less than three gallons remaining in the 12-gallon tank at the end of the day, and our overall average for the 1,711 miles covered was 32.4 m.p.g.

In the small hours of Sunday we were again out in the Place to see the competitors through, and C.D., having wasted half a dozen bulbs in discovering that something was amiss with the flash arrangement on his camera, went back to bed somewhat aggrieved.

The following day's run to Champagnole, where we were to pick up the rally proper on the Monday, was our one leisurely drive and provided an opportunity for the Triumph to show that it can potter pleasantly just as well as motor furiously. The weather was also hot enough for jackets to be discarded and, as occasion demanded, prove once again to ourselves that at least one good thing came from the German occupation of France—a vast improvement in the quality of the beer.

As a place, Champagnole has little to commend it except the cowbells of its cattle, its flat-topped trees and the Grand Hotel G. Ripotot, which is a very pleasant place indeed and offers a particularly succulent *coq au vin* as one of its gastronomic specialities. That evening, while I telephoned a brief report back to *The Motor* office in London, C.D. had another abortive go at his flash mechanism which left him once more in a mood which only a good dinner and several *fines* could ameliorate.

At lunch the following day, before the first competitors were due, we were joined by Pieter Nortier, president of the Dutch "R.A.C. West" which runs the Tulip Rally. With typical thoroughness, and in a Mercedes 300 SL which, he said, was a most exciting car for main road use, but a bit of a handful for motor mountaineering, he was covering the entire route ahead of the competitors to ensure that everything was as planned. Some idea of the amount of time and work which goes into a rally like the Tulip may be gathered from the fact that Nortier reckons he, personally, spends at least seven months of the year doing nothing but organize the next year's event, covering in the process many thousands of miles. It is, no doubt, a result of such attention to detail that the Tulip Rally is among the most smoothly run events of its type anywhere.

And then the first competitors arrived, and off we went in a torrential downpour to cover the southern loop of the combined route embracing Valence, Montelimar (where we couldn't buy any nougat because it was the middle of the night), Grenoble, Chambery, Aix-les-Bains, Gex, in the sort of "no-man's-land" between France and Switzerland, and Belfort.

From our point of view, the rally consisted of staying

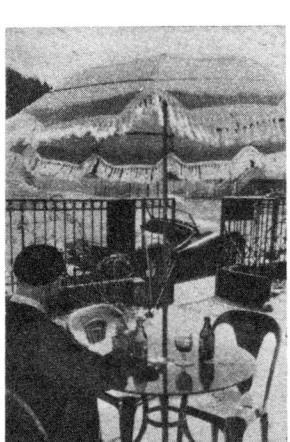

"... a vast improvement in the quality of the beer."

June 22, 1955

From One Generation to Another - Contd.

"... the TR2 suspension is firm enough for all fast motoring needs.... Nor is the physical act of driving at all wearisome."

as long as we dared at one vantage point or another and then making a frantic dash to the next so as not to run out of competitors. Many of our short-cuts, made with the aid of Michelin's excellent 2 km. maps, were over mountain by-ways, hairpin following hairpin, up and down narrow, bumpy roads where the TR2 showed up well and a larger vehicle would have been a time-wasting embarrassment.

The event was fully described in *The Motor* as recently as May 11, so I do not propose to go over it all again here. However, two items, indirectly connected with the rally, stick in my mind, and may be worth recording. The first is our 1.30 a.m. meal in the Auberge du Pin just outside Valence. The timed ascent to this little mountain inn was one of the rally eliminating tests. Except for a few officials and a handful of onlookers, the place was deserted. But, certainly, a meal could be provided, although Madame regretted that there was but a poor choice of viands—one would understand that the inn was not really open. If, however, a little trout, followed by chicken fried in butter, followed by asparagus and then perhaps some fresh strawberries or a little cheese would do, Madame would be delighted. So, indeed, would we—and were, and not the least by the bill which, including coffee and a carafe of quite palatable wine, was, by present French standards, extremely moderate—and all that in the early hours of the morning on top of a minor mountain with Monsieur, Madame and their daughter doing the whole thing themselves. British innkeepers please copy.

The second incident concerned a Danish-entered saloon car which arrived at the finish of the climb up the Col de la Faucille with an anvil chorus sounding in the crankcase. Instead of stopping there and then, the crew continued on down the other side of the pass, at the foot of which we found them, miles from the nearest garage and surrounded by hills. A tow was an obvious necessity. "But it is too small," they protested, pointing at the Triumph. However, we knew our car better than they and finally, with the greatest of ease and at good speed, the damaged car was pulled back to the summit of the Faucille, whence it could free-wheel practically into Geneva. From choice, one would not normally indulge in such hunter-in-the-haycart tactics, but it does serve to illustrate what a maid-of-all-work this Coventry-built two-seater can be when occasion demands.

And so back to England in the same brisk manner as the journey out, including chasing an enterprisingly driven 1900 Alfa Romeo saloon through cloud on a mountain pass and then showing it "the way to go home," all with the same complete absence of driving fatigue, although we had practically lived in the car for five solid days. As C.D. remarked as we garaged the Triumph on the Dunkirk Ferry: "They may call it a sports car, but it's a grand little wagon for *travelling.*"

"... chasing an enterprisingly-driven 1900 Alfa Romeo saloon through cloud on a mountain pass...."

A TR2—PLUS

Derrington modifications add even more performance to an already outstanding car

About the beginning of 1954, I road-tested a new Triumph sports car. At that time, incredible to relate, the cognomen "TR2" meant nothing to the man in the street, and all the many competition successes were still hidden in the future. How time flies!

It's fun to say, "I told you so", and, therefore, it is pleasant to turn back to that old report. The very first sentence reads—"The Triumph TR2 is the most important new sports car which has been introduced for some time". For once, Bolster was right. All of which introduces the subject of the current test, which is a "tuned" version of the same model.

The Triumph TR2, in standard form, gives a genuine 100 m.p.h. performance, coupled with remarkable reliability and fuel economy. That, one might think, is enough to be going on with from a relatively cheap and well-equipped sports car. The answer, of course, is that it's only enough until you want to beat another TR2! These cars appear in club events every week-end, and so a demand has grown up for special equipment that will extract a few more b.h.p. from the willing engine.

V. W. Derrington, of 159-161 London Road, Kingston-on-Thames, has long been known as a purveyor of bits and pieces for the man in search of extra speed. He has now turned his attention to the TR2, and, in fact, races one of these cars himself. I recently borrowed this machine for a week, and these notes are the result.

Bigger Carburetters

The most important modification is a new induction system, with larger carburetters. These are SU instruments, as are the standard ones, but they have a bore of 1¾ ins. instead of the normal 1½ ins. The new inlet manifold has a balance pipe, and blends the larger carburetters with the existing ports, so that only the minimum of "marrying up" is necessary.

In addition, Derrington's car has a new exhaust system. The swept pipes pair off cylinders 1 and 4, 2 and 3, entering the standard silencer via a junction box. An extra straight-through silencer has been fitted in the tail pipe to cure that well-known raucous note.

The induction side of the job, including the two big carburetters, costs £40. The exhaust system comes to £25, with an extra £1 5s. for the tail pipe silencer. Incidentally, the test car also has a Scintilla Vertex magneto, and a few other detail modifications of which more anon.

On the road, the first impression is that the "tuned" car is quieter than the standard model. I like fast cars to be silent, and so I applaud this exhaust system. No loss of flexibility is occasioned by having larger carburetters, and the traffic manners are impeccable. Cold starting is instantaneous.

Standing ¼ mile—17.6 secs.

The acceleration is identical to standard up to 30 m.p.h., wheelspin being the deciding factor in this range. Further up the scale, a steady improvement is recorded. Naturally, a very large power increase would be required to make spectacular gains, but those few useful fifths of a second may make just the difference to beating the other chap into the next corner. I did 0·60 m.p.h. on a re-calibrated speedometer in 10.8 secs., and the standing quarter-mile occupied 17.6 secs.

As regards speed, I got a timed maximum of 107.1 m.p.h. In case you are a Triumph owner, this means 116/118 m.p.h. on the speedometer, if yours is a similar instrument to the one on the test car. These runs were made with the hood and sidescreens in position.

It is needless to remark that one cannot get something for nothing. The improved power output is obtained by passing more air through the engine at maximum revs., mixed with an appropriate quantity of petrol. In the hands of a fast driver, the modified car uses up to 20 per cent. more fuel than a standard one, which is about what one would expect. Apart from this greater thirst, however, there are no other disadvantages.

Modified Bucket Seat

Another useful modification had been made to the driver's bucket seat. Its side had been extended to hold one from slipping towards the passenger's seat, and a padded buffer was also fitted inside the offside door. There are many other makes of cars that would benefit from such treatment, some of them actually being dangerous through the lack of lateral support provided. I am repeatedly telling puzzled drivers that their mysterious steering maladies and cornering difficulties may all be traced to the seat of their pants. I enjoyed flinging Derrington's Triumph through the corners with my posterior firmly anchored in the seat. Would-be competition drivers should certainly check this point.

A common fault among sports cars is loose spokes in their wire wheels. Modern wheels seem particularly prone to this, and the test car has rebuilt rear wheels. These have 64 spokes each, and the 4.50 ins. rims give extra support to the tyres. The battery has also been taken from the bonnet to the boot, which supplies a little more weight to hold down the rear axle at the expense of luggage room.

Very beautiful is the wood rimmed steering wheel. Of laminated and riveted construction, this adds greatly to the appearance of the driving compartment, and is pleasant to the touch. It is the kind of thing that makes a popular model stand out from the crowd, if one has £9 15s. to spare.

The Triumph TR2 is an excellent sports car, and in standard form it will satisfy the majority of owners. For those who want a TR2 with a difference, however, a visit to V. W. Derrington is more than worth while.

WAY IN is through a pair of 1¾ ins. SU carburetters (instead of 1½ ins.) plus a new manifold with balance pipe.

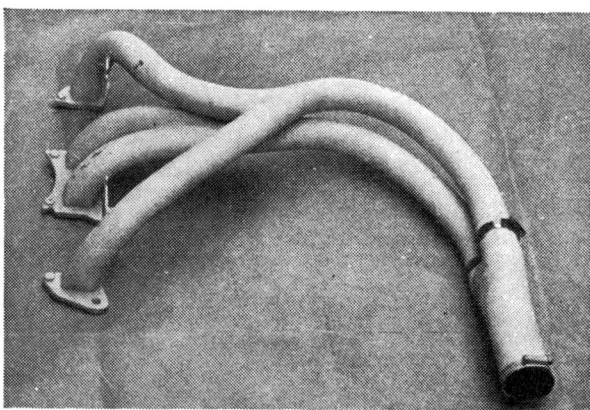

WAY OUT is via this new exhaust system which pairs off cylinders 1 and 4, and 2 and 3. An extra tail-pipe silencer is available.

By AL BRANNON

Trying The Triumph

One of the most popular of the under-$3,000 imports, the TR2 is really finding a place for itself in sports car motoring circles. Popular comment: "It's fun to drive."

TRIUMPH'S TR-2 was one test car that we really hated to give back! In fact, it's been a long time (since the MG TC was new and shining) since we've enjoyed driving a car so much and had one that responded so willingly to the controls.

To begin with, we had heard several irresponsible rumors that the car didn't handle properly, or that it had several weak spots, so we decided to try to find them for ourselves. Triumph went along with the idea and gave us a car to "try to tear up." Our conclusions on the dangerous handling is simply that this little bomb is *really* loaded in the engine department and can easily be overdriven. There isn't a car in the world, with comparable power-to-weight ratio and as husky an amount of torque, that you can't get into trouble if you try.

By the same token, the car is docile as a lamb when driven conservatively and if you're trying to sneak around the streets without attracting too much attention, it can be puttered along at ten miles or so an hour in top gear, and still go nicely when you put your foot in it.

Our testing began, literally, when Paul Thurston of Guylay Associates drove up with the little white car. We piled our photographic equipment into the not-too-large trunk, and headed for the country.

The first thing that impressed us was the healthy bellow from the exhaust and, oh, that wonderful second gear!

It was a great (and only partially withstood) temptation to engage in stoplight dices with some of our more vaunted number-named machinery. We picked a few twisting back roads and tentatively tried a few corners, and by the time we had had three or four back of us, found that the little rocket was covering a heck of a lot of ground at speeds that we normally can't manage.

By the time we got to our test grounds, the car felt familiar and we pushed it around for about an hour experimenting with the cornering. Here we found a couple of interesting points about the handling characteristics. To drift, the TR-2 requires a relatively high speed and a pretty violent cut at the wheel, coupled with some conservative work on the throttle. The back end is a bit light and if you let it, it will break away quite rapidly. Low-speed corners can be managed in almost midget style, with a series of little skids, controlled by the throttle. We found that this technique will get you around a bend almost as fast as a drift, and the power in reserve is enough to literally blast you out of the corner on the other side.

Reluctantly, when it began getting dark, we decided to call an end to our playing for the day and headed for home. Unfortunately, the wife had heard the car approaching and said something to the effect that, "I like it!" We then waited an hour or so for her to return from "taking it around the block." Believing that a woman's opinion

An enthusiast test of the TR2

Front end is clean and functional—although some may feel it's a bit too stark.

Front suspension is indicative of the sturdy construction of this sports car.

of the car would be interesting, we tried to get some usable statement from her on the machine. "I like it!" seemed to be the order of the day, so we decided that that could be the summary. Seriously, though, the car will handle like a baby carriage when driven conservatively and shouldn't prove to be too much of a handful for anyone not bent on establishing records to and from the corner grocery.

The second day we had the car, we also had thunderstorms. This gave us an opportunity to try the handling in the wet and also to experiment with the top and side curtains. The handling was good and seemed to be no more skittish than on dry pavement. The top goes up, not without a bit of perseverance, and the side curtains really seal things up tight. In fact, they're too tight for summer rain storms, causing the windshield to mist pretty badly. We tried cutting off the hot water to the heater and found that the fan makes a workable de-mister. With the top up, though, an outside mirror is a must.

The following day we tried some acceleration runs and found that the car would consistently do 0-30 in 3.5 seconds, 0-40 in 5.7, 0-50 in 8.1, and 0-60 in 11.8. We tried running over the red line to 60 a couple of times, and in spite of a very slight uphill grade, managed to reach the mark in 10.5 seconds both times. In these runs we were somewhat appalled at all the power going into wheel spin instead of acceleration, and feel that the car would not be handicapped by adding a couple of hundred pounds to rear end. In low gear, if you try, the rear wheels will spin long enough for you to wonder when they're going to stop.

Our top speed runs were made with the top and side curtains in place and we arrived at 95 mph. with some unused pedal left over. Our car wasn't equipped with overdrive, but we believe that somewhere between 100 and 105 would be about right for the machine.

Once through with the calibrations, we decided to try the car on handling again, this time with a couple of other test drivers. We put it through corners every wrong way we could imagine and not once did it feel unstable. True, we managed some rather nice spins, but nothing that could not be avoided with only half a try. We *didn't* manage to induce any brake fade, or any difficulties with the hubs.

However, we found some things about the car that we didn't like. For our height, the front stake pocket affair that holds the left-hand side curtain left quite a collection of bruises on our knee cap, and the steering wheel couldn't be moved close enough to the dash for adequate arm room

without causing quite a reach for the pedals. Of course, for drivers who like the wheel close to them, this won't be objectionable.

We weren't impressed with the finish of the car. It is obviously built to a price and the fit of the hood and trunk lids left something to be desired. Speaking of appearance, that front end is still a little too stark. True, it's functional, inasmuch as it is a hole to let air into the engine, and that's exactly what it looks like, but—

On the rest of the car, the instrumentation is wonderful and the positioning one of the best we've seen. The gearbox is good and the shift short and very positive. On the other hand, we'd like to feel a little more of the 'works' through the clutch.

As to the weather equipment, it's good. The drawback is the little fasteners. They hold things together well, *too* well in fact, and sometimes require pliers to release them. The rubber apron material of which the top and side curtains are made, looks like it would stand up indefinitely and does make a neat installation when in place.

As a summary, it was the opinion of everyone who drove the car that it was, if anything, too potent. It is a car that should be able to hold its head up in competition with any of the two-liter class and its price is way below most of its class.

In addition to the advantage of the low initial cost, Triumph has formed an owners association to assist in clearing information on competing the cars, tuning and a listing of the cars' successes. From the looks of the book, it should be a regular gold mine.

We did conclude that the car, although it makes you feel at home right from the start, is definitely not an amateur's machine as far as racing goes. It is potent, and should be given at least as much time in practice as any other 2,000 cc. sports-racing car. As we have said before, you can overdrive any car and one as hot as the TR-2 is no exception.

All in all, the car is a 'happy' one that is downright fun to drive and if the newcomer will treat that potent second gear with respect until he's thoroughly accustomed to the feel of the machinery, it will give a lot of the bigger, more expensive machines a real run for their money and in a lot of instances flatly outdo them.

Specifications: Wheelbase, 88 inches; tread (front), 45 inches; tread (rear) 45½ inches; ground clearance (under axle), 6 inches; turning circle, 32 feet; tire size, 5.50x15; length, 151 inches; width, 55½ inches; height, top up, 50 inches; top of windshield, 46 inches; windshield removed, 40 inches; number of cylinders, 4; bore, 83 mm.; stroke, 92 mm.; capacity, 1,991 cc.; compression ratio, 8.5; firing order, 1,3,4,2; bhp., 90 at 4,800 rpm.; torque, 1,400 at 3,000 rpm.; bmep, 145 lbs./sq.in.; fuel consumption, 32 mpg.; oil consumption, 3,000 mpg.

—☆—

Positioning of the instrumentation on the dash leaves little to be desired. And the shift is short and very positive.

The Motor Road Test No. 5/56 (Continental)

Make: Triumph **Type:** TR3 Hard-top Coupé (with overdrive)
Makers: The Triumph Motor Co. (1945) Ltd., Coventry

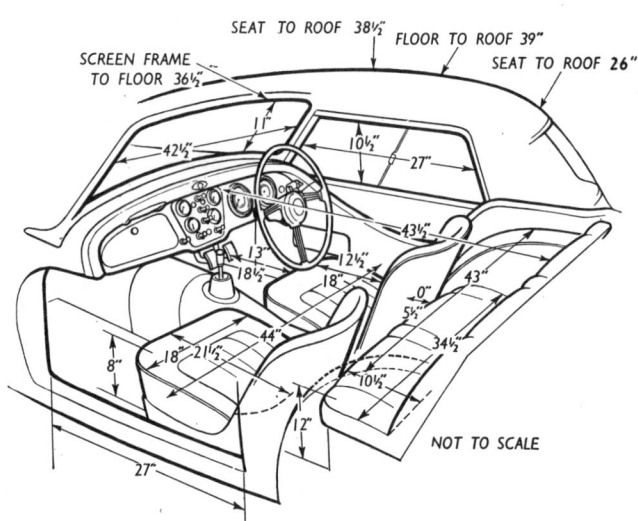

Test Data

CONDITIONS. Cool, dry weather with little wind. (Temperature 32-40°F., Barometer 30.2-30.4 in. Hg.). Smooth concrete road surface (Ostend-Ghent motor road). Premium grade Belgian pump fuel. Tyre pressures raised 6 lb./sq. in. as advised for fast driving.

INSTRUMENTS
Speedometer at 30 m.p.h.	Accurate
Speedometer at 60 m.p.h.	5% fast
Speedometer at 90 m.p.h.	2% fast
Distance recorder	Accurate

MAXIMUM SPEEDS
Flying Half Mile (Direct Top Gear)
Mean of four opposite runs ... 105.3 m.p.h.
Best time equals ... 105.9 m.p.h.
Flying Half Mile (Overdrive Top Gear)
Mean of four opposite runs ... 104.7 m.p.h.
Best time equals ... 106.5 m.p.h.
Speed in Gears (at 5,000 r.p.m. recommended limit)
3rd gear, 76 m.p.h. (overdrive, 94 m.p.h.)
2nd gear, 51 m.p.h. (overdrive, 62 m.p.h.)

FUEL CONSUMPTION (direct top gear)
43.5 m.p.g. at constant 30 m.p.h.
41.5 m.p.g. at constant 40 m.p.h.
38.0 m.p.g. at constant 50 m.p.h.

FUEL CONSUMPTION (overdrive)
45.5 m.p.g. at constant 40 m.p.h.
42.0 m.p.g. at constant 50 m.p.h.
37.5 m.p.g. at constant 60 m.p.h.
33.0 m.p.g. at constant 70 m.p.h.
29.0 m.p.g. at constant 80 m.p.h.
25.0 m.p.g. at constant 90 m.p.h.
Overall consumption for 2,646 miles, 97.7 gallons, =27.1 m.p.g. (10.4 litres/100 km.).
Fuel tank capacity, 12½ gallons.

ACCELERATION TIMES Through Gears
0-30 m.p.h.	3.6 sec.
0-40 m.p.h.	5.4 sec.
0-50 m.p.h.	7.5 sec.
0-60 m.p.h.	10.8 sec.
0-70 m.p.h.	14.6 sec.
0-80 m.p.h.	20.2 sec.
0-90 m.p.h.	28.8 sec.
Standing Quarter Mile	18.1 sec.

ACCELERATION TIMES on Upper Ratios
	Overdrive Top	Direct Top	Direct 3rd
10-30 m.p.h.	—	9.2 sec.	6.2 sec.
20-40 m.p.h.	10.8 sec.	8.6 sec.	5.7 sec.
30-50 m.p.h.	11.5 sec.	8.9 sec.	6.0 sec.
40-60 m.p.h.	12.4 sec.	9.1 sec.	6.3 sec.
50-70 m.p.h.	13.7 sec.	9.7 sec.	7.0 sec.
60-80 m.p.h.	15.2 sec.	11.3 sec.	—
70-90 m.p.h.	19.5 sec.	14.4 sec.	—

HILL CLIMBING (at steady speeds)
Max. gradient on overdrive top gear ... 1 in 11.1 (Tapley 200 lb./ton)
Max. gradient on direct top gear ... 1 in 8.4 (Tapley 265 lb./ton)
Max. gradient on overdrive 3rd gear ... 1 in 7.8 (Tapley 285 lb./ton)
Max. gradient on direct 3rd gear ... 1 in 6.2 (Tapley 355 lb./ton)
Max. gradient on overdrive 2nd gear ... 1 in 4.6 (Tapley 475 lb./ton)
Max. gradient on direct 2nd gear ... 1 in 3.8 (Tapley 565 lb./ton)

BRAKES at 30 m.p.h.
0.94g retardation ... (=32 ft. stopping distance) with 100 lb. pedal pressure
0.77g retardation ... (=39 ft. stopping distance) with 75 lb. pedal pressure
0.44g retardation ... (=68 ft. stopping distance) with 50 lb. pedal pressure
0.24g retardation ... (=126 ft. stopping distance) with 25 lb. pedal pressure

WEIGHT
Unladen kerb weight ... 19 cwt.
Front/rear weight distribution ... 53/47
Weight laden as tested ... 22¼ cwt.

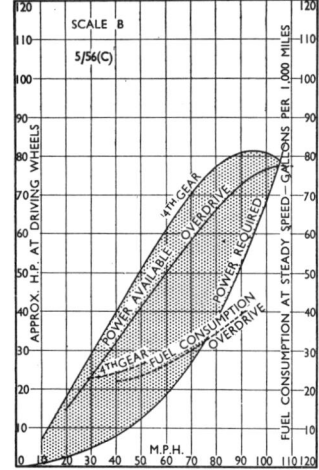

Drag at 10 m.p.h. ... 39 lb.
Drag at 60 m.p.h. ... 115 lb.
Specific Fuel Consumption when cruising at 80% of maximum speed (i.e. 84.2 m.p.h.) on level road, based on power delivered to rear wheels ... 0.57 pints/b.h.p./hr

Maintenance

Sump: 10 pints (plus 1 pint for filter), S.A.E. 30 (summer), 20 (winter). **Gearbox:** 1½ pints (plus 2 pints if overdrive fitted), S.A.E. 30. **Rear axle:** 1½ pints, S.A.E. 90 E.P. **Steering gear:** S.A.E. 90 E.P. **Radiator:** 13 pints (plus 1 pint if heater fitted) (2 drain taps). **Chassis lubrication:** By grease gun every 1,000 miles to 13 points, and every 5,000 miles to 13 additional points. **Ignition timing:** 4° B.T.D.C. **Spark plug gap:** 0.032 in. **Contact breaker gap:** 0.015 in. **Valve timing** (set with 0.015 in. clearance): Inlet opens 15° B.T.D.C. and closes 55° A.B.D.C. Exhaust opens 55° B.B.D.C. and closes 15° A.T.D.C. **Tappet clearances:** (Cold). Inlet 0.010 in. Exhaust 0.012 in. (for high-speed running 0.013 in. for inlet and exhaust). **Front wheel alignment:** Parallel to 1/16 in. toe out. **Camber angle:** 2°. **Caster angle:** 0°. **Tyre pressures:** Front 22 lb., rear 24 lb. (increase both by 6 lb. for sustained high speeds). **Brake fluid:** Lockheed orange. **Battery:** 51 amp. hr.

The TRIUMPH TR3 Hard-top Coupé
(with overdrive)

Popular 2-litre Sports Model with Improved Acceleration and Amenities

OPTIONAL equipment for the Triumph TR3, here seen above Switzerland's Rhone Valley, is this bolt-on steel roof incorporating a curved-glass rear window. Removable sidescreens have sliding panels for ventilation, and canvas flaps held closed by lift-the-dot fasteners.

SINCE we tested the Triumph TR2 two-seater in the spring of 1954, what was then a striking new model has been developed in two distinct directions. The basic design has been improved to a very great extent by many inconspicuous changes, and the appeal of the model has been broadened by introduction of a variety of extras. Basically, there is in the latest Triumph TR3 more power than formerly, this giving usefully improved acceleration through the gears although maximum speed remains little altered. Many other items which called for criticism in our 1954 Road Test Report have been visibly improved, such as the seats and the doors, whilst the chassis also seems to behave better at speed on rough or slippery roads.

Additionally, the virtues of a fast-moving and economical sports car have been made available to a wider public, by provision as optional features of a bolt-on steel roof incorporating a big curved-glass rear window, and of a miniature rear seat suitable for two young children, both these extras being fitted to the model tested. Other extras fitted to the test car were the Laycock-de Normanville overdrive, an interior heater and de-mister of recirculating type, and leather upholstery.

Improved amenities and extras have added only ½ cwt. to the Triumph, which on ordinary tyres recorded a maximum speed of over 105 m.p.h., and 0-50 m.p.h. and 0-90 m.p.h. acceleration figures of 7.5 sec. and 28.8 sec. An overall fuel consumption of 27.1 m.p.g. for more than 2,500 fast miles, with the quite notable high-cruising speed economy of 33.0 m.p.g. at 70 m.p.h. in overdrive, is noteworthy.

The increased output of the power unit has in no way detracted from its behaviour under touring conditions. This engine is, of course, a developed edition of the well-known Vanguard unit and, as such, it has a four-cylinder feel at low speeds. It is, however, remarkably flexible for a sports car power unit. On normal British premium petrol, virtually no pinking is in evidence, but the compression ratio has obviously been chosen to obtain maximum benefit from present British No. 1 grades, because the engine pinks quite loudly with the slightly inferior premium fuels sold in some Continental countries, especially at low speeds in traffic when hot.

A point on which the car can very definitely be criticized, however, is on the score of exhaust noise which is still far too loud at speeds of 2,000 r.p.m. upwards—a point which not only tends to make a driver restrict full use of the performance on occasion but is also liable to bring motor sport into disrepute by causing annoyance during all-night rallies. Also, at low r.p.m. there is a certain amount of mechanical noise, and at much above 70 m.p.h. wind noise make conversation tiring. Notably good features are easy starting and quick warming up, and the first-rate accessibility of all components needing routine attention.

Except for one comparatively minor detail, the transmission arrangements of the TR3 inspire enthusiasm in even the most blasé driver. The central gear lever is of the stubby type, which is placed exactly where the hand falls on it naturally, works in a commendably positive fashion and has a delightfully short travel, although on the test car the change from 3rd down to 2nd gear was not always smooth. In conjunction with a clutch which is appropriately positive for a sports car but still smooth enough for second gear starts if a moderate degree of finesse is used, the gear change is ideal for

---

In Brief

Price (Hard-top model with overdrive): £735 plus purchase tax £368 17s. equals £1,103 17s.
Capacity ... ... 1,991 c.c.
Unladen kerb weight ... 19 cwt.
Fuel consumption, driven fast 27.1 m.p.g.
Maximum speed ... ... 105.3 m.p.h.
Maximum top gear gradient 1 in 8.4
Acceleration:
 10-30 m.p.h. in top ... 9.2 sec.
 0-50 m.p.h. through gears 7.5 sec.
Gearing: 20.2 m.p.h. in direct top at 1,000 r.p.m. (24.6 m.p.h. in overdrive top); 33.4 m.p.h. in direct top at 1,000 ft. per min. piston speed (40.8 m.p.h. in overdrive top).

SOME PROTECTION for the tail is provided by vertical over-riders instead of by a bumper bar, the panel between them removing to give access to the spare wheel and tools. Ground clearance is moderate, but the underside of the car is flat.

The Triumph TR3 Hard-top Coupé

both rally-type tests and for getting the best out of the car under normal conditions with the least effort.

If the buyer elects to have the Laycock-de Normanville overdrive, an additional three ratios are at his disposal. The undoubted merits of the Laycock-de Normanville device in providing both effortless cruising and fuel economy in top gear are too well-known to need stressing here. What might be considered more debatable is the advantage of overdrive as applied to the third and second gears of a four-speed box. In this matter, much depends on how the ratios fall and, in the case of the TR3, which has an unusually close-ratio set of overdrive gears, there is no useless overlapping between overdrive second and normal third or between overdrive third and normal top, so that all the seven ratios are, in effect, separate and distinct. In normal motoring, it is quite unnecessary to exploit the subtleties of these seven speeds to the full, but for competition work the availability of an exactly appropriate ratio can be of the greatest benefit.

The overdrive control is particularly handy, taking the form of a neat switch on the offside of the facia panel, where it can be operated by two fingers without the right hand leaving the wheel. This makes it a simple matter to effect simultaneous double changes such as switching from overdrive to direct whilst operating the normal gear lever. The only point on which the overdrive is not ideal is that engagement (on the car tried at least) was definitely harsh at many speeds even when, in accordance with the correct procedure, the throttle was kept open during the change, although downward changes into direct drive were very smooth. A considerate driver would generally use the clutch for upward changes on the test car, except when out for maximum acceleration. Even without the overdrive, 80 m.p.h. seems a very reasonable cruising speed on suitable roads.

Well in keeping with the performance of the car are the brakes, which provide good stopping powers with quite moderate pressures, but a minor annoyance was an occasional tendency to grab at low speeds, especially after the car had been standing. The handbrake is of the racing fly-off type, powerful and nicely placed in relation to the gear lever although apt to rub the driver's left leg. There is plenty of room for the left foot to the side of the clutch pedal, where the dipper switch acts also as a foot-rest, and pedal spacing permits driving in broad shoes.

In most other respects, the controls and instruments are very well placed indeed. Everything is just where it is required, including the rev. counter and speedometer with large dials and clear faces just in front of the driver, the designers having also avoided the common elementary mistake of making the speedometer the right-hand of the two where it cannot be seen by the passenger. Minor faults are slightly over-bright instrument illumination and the use of three identical switches one above the other on the central panel, which makes it all too easy to operate the wipers in mistake for the lamps.

Quick Steering

Steering and handling qualities are satisfactorily sporting. The gearing of the former is high, and whilst this contributes to slightly heavy steering at low speeds, it is unobjectionable for a sports model. There is a moderate degree of understeer which increases on wet road surfaces, and very slight roll on corners; at high speeds the car follows a straight course with very little attention to the wheel and is not oversensitive to cross winds. The springing itself gives good bump absorption without undue movement at speed, and the car rides *pavé* well with the tyres at the lower of the two suggested pressures.

OCCASIONAL seating which may be provided in the back of the Triumph TR3 is designed for young children. Bucket-type front seats are individually adjustable.

VALUABLE capacity for luggage under lock-and-key is provided in the square-cut tail of the body, on a flat floor with no obstruction by spare wheel or tools.

Aiding the comfort and good control are separate front seats which have well-padded cushions and nicely-rounded squabs, which give good side support although the thinly upholstered squabs slope backwards a little too much for the tastes of some. Inevitably, a low car is somewhat hard on the heels of the occupant's shoes, but in this case the rubber floor covering was also showing surprisingly rapid wear. Internal width across the body, incidentally, has been increased by some 3 in. which adds considerably to the comfort of the car.

Passing judgment on the new occasional rear seating is more difficult. For two very small children travelling behind parents of moderate height the arrangement is useful. For a larger occupant, such as a teenager travelling behind a front pair who require their seats in more rearward positions, the occasional seating is of very little use because knee room is virtually non-existent, and a single transverse seat might have been of more general use.

The TR3 offers a clear view to the front (both wings and headlamp cowls are visible) and the wrap-round rear window is excellent for reversing. The view to the side is also good, but the extremely low

The Motor ROAD TESTS OF 1956 CARS

— — — — — — — Contd.

roof restricts upward vision in this direction to some extent, and six-foot occupants continually brush the headlining with their hair.

On the whole, the optional hard-top may find favour over the normal hood, for those who use their cars in closed trim almost all the year round, but it has positive disadvantages in space and noise without any great improvement in weather protection. Back draughts are notoriously brackets which fit into taper slots with a screw locking device. In warm weather, the car is quite pleasant without them and they can be accommodated in the boot.

The latter is of generous size, for a car of this price and performance, and will accommodate one very large suitcase or two small suitcases plus a surprising number of odds and ends. There is also space behind the front seats for a second large suitcase if the rear seat is not in use. For oddments, there are a very useful lockable cubby, and good pockets in the doors.

ENLARGED carburetters which are fitted to the latest Triumph engine shift the emphasis from economy to performance. Most under-bonnet components are readily accessible.

difficult to avoid on small closed cars and it cannot be said that the Triumph designers have succeeded entirely in this respect. The re-circulating heater is very adequate and has the merit of having small doors by which the supply of hot air can be concentrated on to the driver's or passenger's legs, or on the windscreen for de-misting, as desired. Like most open cars, this model lets in some water here and there on very wet days.

The new sidescreens are amongst the best we have encountered, with neat sliding windows of transparent plastic. They are commendably rigid, a snug fit and attached by robust wedge-shaped support

It is perhaps in these features which make it so extraordinarily suitable for day-to-day motoring of every kind that the Triumph is really striking. Performance comparable with that of almost any car of less than twice its price is combined with roadworthiness of at least a good average sports-car standard, and unusual economy. Its outstanding virtue, however, is the practical layout which makes it possible for a man to run this as his only car, rather than an occasional and delightful toy. Usefulness and entertainment in a motorcar do not often come so close together.

Mechanical Specification

Engine
Cylinders ... 4
Bore ... 83 mm.
Stroke ... 92 mm.
Cubic capacity ... 1,991 c.c.
Piston area ... 33.5 sq. in.
Valves ... Push-rod o.h.v.
Compression ratio ... 8.5/1
Max. power 100 b.h.p. at 5,000 r.p.m.
Piston speed at max. b.h.p. 3,010 ft. per min.
Carburetters ... Two inclined S.U. (H.6)
Ignition ... 12-volt coil
Sparking plugs ... Champion L 10 S
Fuel pump ... AC mechanical
Oil filter ... Purolator by-pass

Transmission
Clutch ... Borg & Beck, 9 in. s.d.p.
Top gear (s/m) ... 3.7 (overdrive 3.03)
3rd gear (s/m) ... 4.9 (overdrive 4.02)
2nd gear (s/m) ... 7.4 (overdrive 6.07)
1st gear ... 12.5
Propeller shaft ... Hardy Spicer, open
Final drive ... Hypoid bevel
Top gear m.p.h. at 1,000 r.p.m. 20.2 (overdrive 24.6)
Top gear m.p.h. at 1,000 ft./min. piston speed ... 33.4 (overdrive 40.8)

Chassis
Brakes Lockheed hydraulic (2LS on front)
Brake drum diameter ... 10 in
Friction lining area ... 175 sq. in.
Suspension: Front Coil and wishbone i.f.s.
Rear... Semi-elliptic
Shock absorbers: Front Telescopic hydraulic
Rear Piston-type hydraulic
Tyres ... Dunlop 5.50—15

Steering
Steering gear ... Cam and lever
Turning circle between kerbs:
Left, 33 ft. Right, 31½ ft.
Turns of steering wheel, lock to lock ... 2¼

Performance factors (at laden weight as tested)
Piston area, sq. in. per ton ... 29.8
Brake lining area, sq. in. per ton 155
Specific displacement, litres per ton mile... 2,650 (overdrive 2,150)
Described in The Motor, October 19, 1955.

Coachwork and Equipment

Bumper height with car unladen:
Front (max.) 18 in., (min.) 9 in.
Rear (max.) 21½ in., (min.) 11½ in.
Starting handle ... Yes
Battery mounting ... On scuttle
Jack ... Screw pillar type
Jacking points Two on chassis, accessible through trapdoors in floor
Standard tool kit: Jack, wheelbrace (or copper hammer with optional wire wheels), sparking-plug spanner, three open-ended and one tube spanner, tyre valve tool, tyre levers, grease gun, tommy bar, screwdriver, nave-plate remover, slip joint grips, adjustable spanner, plastic tool roll.
Exterior lights: Two F700 headlamps, two side/flasher lamps, two rear/flasher lamps, combined number plate/stop lamp.
Direction indicators Flashers, self-cancelling
Windscreen wipers ... Two-blade electric, non-self-parking
Sun vizors ... None
Instruments: Speedometer (with decimal trip), rev. counter, fuel gauge, oil pressure gauge, ammeter, water thermometer.
Warning lights Ignition, direction indicators, headlamp main beam
Locks: With ignition key ... Ignition
With other key Cubby locker and boot
Glove lockers ... One (with lock)
Map pockets ... Two (in doors)
Parcel shelves ... None
Ashtrays ... None
Cigar lighters ... None
Interior lights ... None
Interior heater ... Optional extra, re-circulating type with de-misters
Car radio ... Optional extra
Extras available: Fabric hood and frame, aero screens, Laycock-de Normanville overdrive, short front undershield, knock-on wire wheels, rear wing spats, leather upholstery, tonneau cover, Trafalgar screen spray, telescopic steering wheel, Road Speed tyres, aluminium sump, radio, two-speed screen wiper, competition front springs, competition rear dampers, fitted suitcase, "occasional" rear passengers seat.
Upholstery material Vynide (leather extra)
Floor covering Rubber in front, carpet at rear
Number of exterior colours standardized: Nine combinations for body and hard-top.
Alternative body styles ... Open model (i.e. without hard top)

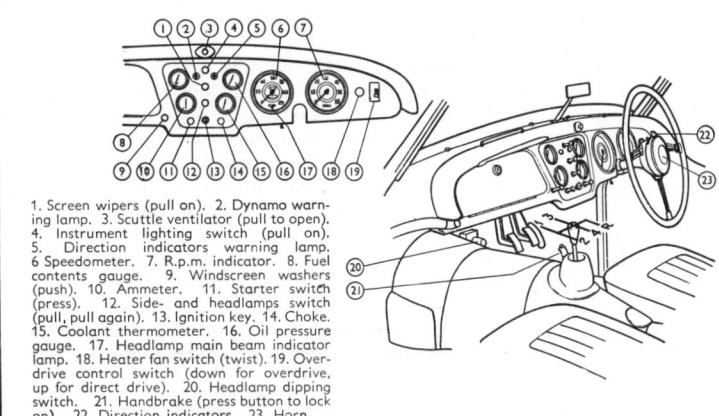

1. Screen wipers (pull on). 2. Dynamo warning lamp. 3. Scuttle ventilator (pull to open). 4. Instrument lighting switch (pull on). 5. Direction indicators warning lamp. 6 Speedometer. 7. R.p.m. indicator. 8. Fuel contents gauge. 9. Windscreen washers (push). 10. Ammeter. 11. Starter switch (press). 12. Side- and headlamps switch (pull, pull again). 13. Ignition key. 14. Choke. 15. Coolant thermometer. 16. Oil pressure gauge. 17. Headlamp main beam indicator lamp. 18. Heater fan switch (twist). 19. Overdrive control switch (down for overdrive, up for direct drive). 20. Headlamp dipping switch. 21. Handbrake (press button to lock on). 22. Direction indicators. 23. Horn.

Newly modified and sporting many changes,

Triumph gets a going over

from headlight to tail bumper in a...

test of the TR3

By KARL LUDVIGSEN

EVER drive a Le Mans finisher? Ever boot around corners in a Tourist Trophy and Alpine Rally Team Prize winner? Or could your car take seventh in its class in the Mille Miglia? For a hot sports/racing car these achievements would be creditable, but when they are recorded by a fully-trimmed sports car, mass-produced to a definite price, they appear startling indeed.

The TR series of Triumph sports cars was born in the summer of 1952, when Sir John Black of Standard Motors paid a visit to the United States. He returned with a target for his associates: a fast yet cheap sports car for that specific market. They first came up with the 75 bhp, exposed-spare TR1, which was subsequently modified over a year of development work and came on the market as the TR2. Early U. S. enthusiasm was slightly tempered by delayed deliveries, but the car caught on quickly, particularly on the West Coast. The factory took up competition work after private owners had scored some successes, carrying on their development work within this sphere.

Knowing that some alterations would boost their chances in the American market, Standard introduced the Triumph TR3 at the 1955 London Motor Show, to supplement the TR2 in their range. External changes include a new "surface" grille, mounted over the old one, and stainless steel fender beading. A back seat for the children is now also available, as is an overdrive that operates on the top three

New grille, features a flush front of rugged vertical bars and smaller horizontal "stiffeners."

speeds of the four speed gearbox. Under the hood, 1½ inch S. U. carburetors have been replaced by the 1¾ inch size, and the power output has thereby been raised from 90 bhp to 100 bhp, the revs also going from 4,800 to 5,000. Interior ventilation has been improved by means of a cowl vent and sliding panels in the side curtains, and a steel hardtop is also available.

The Standard-Triumph Motor Company, Inc. accepted the risk of letting me try their first TR3, and I was able to get a pretty good impression of it over some 350 fast miles. First, I might mention that the TR3 retails for $100 more than the TR2, putting the basic price at $2599. "My" car had the hardtop, wire wheels, a heater, adjustable steering, the rear seat, and overdrive, which combined would put the F.O.B. port price of this car at $3180.80. The original TR2 was built strictly to a price limit of $2500, but the deluxe TR3 actually steps into quite a different class. You must, of course, evaluate the various accessories according to your purposes, but I hope I can give you something to go on.

Looking over the general body layout, the bolt-on fenders should greatly ease repairs, as will the broad hood opening. Engine accessibility is the best of the well-known sports cars, with all components very easy to reach. Only the steering box and the generator might be out of the way. The spare tire, jack, etc. have a housing of their own at the rear, leaving the wide and deep luggage compartment free. All these openings can be unlocked only with a special T-handled tool, while the trunk has an additional key lock.

The doors are opened from the outside by sliding open the side window and pulling up on the cord-controlled latch, which works smoothly if kept well greased. A wide opening makes entry as easy as possible in such a compact car, and, once inside, I found more than ample room for six feet of height. Leg and foot room are truly outstanding, there being considerable width between the gearbox and the left wall. My left foot rested naturally on the dimmer switch, while the right could use a bit more support at the pad-type accelerator. The seats themselves are better cushioned than the early cars, and provide firm and upright support. The backs are well contoured, but do not hold the hips laterally.

Adjustable on this car, the 17 inch steering wheel is well positioned and does not conflict with the thighs. If you like the wheel at a distance, "Italian style," get the non-adjustable model, for it is closest to the dash of all. Instrumentation is complete, impressive, and legible. All knobs are easy to handle, though the choke wouldn't always stay at the desired position. The overdrive switch protrudes several inches from the dash on the left, and moves up for direct and down for overdrive. This always seemed illogical to me. A key is necessary to open the roomy glove compartment, and draft-free door pockets will hold more incidentals.

When the side curtains are out there is plenty of elbow room, but when they are replaced they restrict left arm action. The rear curtain holder is a particular elbow-knocker. I had the TR3 out in near-freezing weather, which provided a severe test of the weather sealing. Conclusion: It is not possible to recommend the optional hardtop as a substitute for a coupe. The car is subject to the same front and rear drafts encountered with the soft top, and in this particular car my passenger fared better than I. The heater and defroster are as potent as any sports car can offer, but can't quite hold their own at speed.

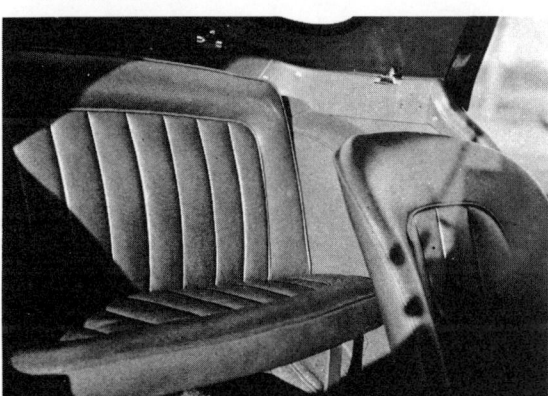

Jump seat is small but can hold extra luggage or a child for short trips. Seat available at additional cost.

TR3 dash has grouped instruments. Tachometer and speedometer are on left and engine instruments on right.

Eight individual bolts hold down the hardtop, which takes it out of the quick-detachable league, but in return eliminates all rattles and greatly stiffens the car as a whole. An attractive interior flocking beautifies but does not soften the steel, and I must record three occasions on which I forgot to duck while negotiating bumps. Headroom is rather restricted in the optional rear seat, where legroom is nonexistent if the front seats are at the rear of their long travel. Triumph sensibly admits that this seat is for small children only, and it would indeed be useful for that purpose. It is very well finished, and enhances the already well-trimmed interior.

Having had a look around, let's switch on and press the starter. Full choking will give a clean start from cold, and it is particularly desirable to warm up the TR3 before setting out. At any temperature below 160°F the engine is prone to cut out when accelerated from below 2000 rpm. It was not at all flexible until warm, and I tended to think of the difficulties of cold vaporization in the new larger carburetor venturis.

Most of my tests were carried out with the engine at 140°F, which is too cold for best results. Even so, the TR3 does not drag its feet. The engine winds fast in a smooth and purposeful manner, reaching a useful peak of around 5400 rpm. The sharp exhaust note of the TR2's has gone, but the car still belts away from a standing start with a real kick-in-the-back. The clutch on this demonstrator was smooth but not as sharp as it might have been when engaged hydraulically by its suspended pedal.

The gearbox is a delight to handle, through its central stubby, short-throw lever. The synchromesh is just about right, and can be overridden by a fast change. One must lift the lever to go into reverse, but this latch can be beaten by a vigorous move from second to third. The rubber knob is comfortable, but to me has a non-mechanical feel.

I was most interested in trying out the Laycock-de Normanville overdrive, which turned out to be a highly useful unit. It shifts virtually instantaneously at a flick of the dash switch, and actually makes its smoothest upshifts under full power, without a trace of slip. A very useful selection of gears is provided, and only third overdrive and direct fourth are at all alike. A couple of times this unit shifted itself to direct, but since it later provided overdrive again I tend to blame a connection somewhere. This automatic control can be used either to get the most out of the car, or to take it easy in town driving. Whatever you like, it complements the gearbox very well. The whole drive train and engine combination, in fact, is very good, though the rear axle tends to wind up a bit from a standing start.

In a car that moves up to the eighties and nineties with such ease, brakes are obviously very important. The TR3 always stopped quickly and in a straight line, without a trace of harshness. I never made them fade in much hard driving, in this wire-wheeled model.

Rallies demand much of the handling and agility of a sports car, and the Triumph has long been tops in that field. The steering itself is reasonably fast, at 2½ turns lock to lock, and includes but an inch or so of free play at the rim. Freed from the shaking that plagued earlier cars, the steering now has a strong self-centering action. When the TR3 is being cornered fast it assumes and holds an initial roll angle, which affects the driver much more than the car's handling.

The Triumph exhibits strong understeer, which means that the faster you take a given bend (say left), the farther left you must turn the wheel to hold the same line. I tried it first with the touring tire pressures of 25 rear and 22 front, and then added five pounds to each. The lower pressures definitely aggravated the understeer. During these maneuvers the rear end moves gradually out in an entirely predictable and controllable manner. The rear shocks have al-

(Continued on page 56)

Hardtop is firmly attached and adds to both rigidity and comfort of the new TR3.

Photographs by Don Typond

TRIUMPH TR3
SPECIFICATIONS
ENGINE
Cylinders 4
Bore & Stroke .... 83 mm x 92 mm (3.27 in. x 3.62 in.)
Displacement ....1991 cc (121.5 cu. in.)
Compression ratio 8.5:1
Max. Horsepower ....100 bhp @ 5000 rpm
Max. Torque ....118 lb. in. @ 3000 rpm
Max. b.m.e.p. ....145 psi
CHASSIS
Wheelbase ....88 in.
Front track ....45 in.
Rear track ....45.5 in.
Dry weight ....2000 lbs.
Test weight ....2500 lbs.
Turns lock to lock 2.5
Turning circle ....34 ft.

Ratios	Overall	Overdrive
4th	3.7	3.03
3rd	4.9	4.02
2nd	7.4	6.07
1st.	12.5	
Rev.	15.8	

Tire size ....5.50 x 15
Brake lining area ....175 sq. in.
Fuel capacity ....14.41 gal.

TRIUMPH TR3
PERFORMANCE
TEST CONDITIONS
40° F, no wind, dry concrete surface at sea level.
SPEEDS IN GEARS   true (car) mph
1st 30 (31)
2nd 53 (55)
2nd OD 66 (68)
3rd 81 (84)
3rd OD ....98 (102) @ 5200 rpm
4th ....101 (105) @ 4750 rpm
Best run 103.5
4th OD ....100 (104) @ 3800 rpm
ACCELERATION   Gears Used
0-30 ....3.8 sec.............1st
0-60 ....11.4 sec...........1st, 2nd, 2nd OD
0-80 ....24.0 sec...........1st, 2nd, 3rd
50-70 ....7.6 sec.............3rd
Standing ¼ mile ....17.8 sec......1st, 2nd, 3rd
Speed at end of quarter ....74 mph

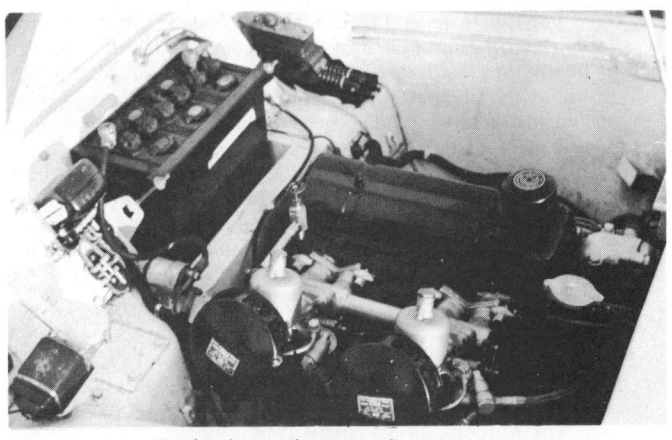

Engine layout is neat and accessible. Ten more horsepower has been added for a bit more "go".

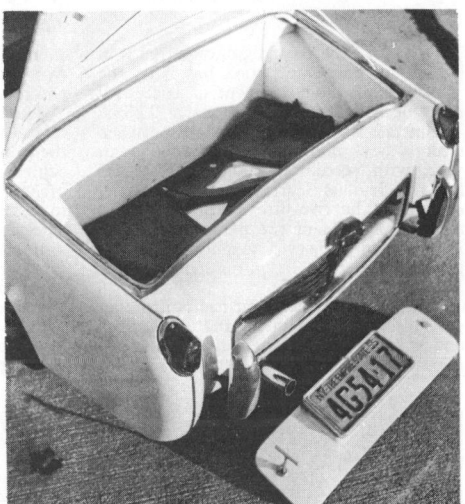

Tire is stored in a compartment below the trunk. Tail light illuminates area for tire changes at night.

Continued from page 54

ways been a Triumph weakness, but the TR3 seems somewhat better in this department. The car holds position well in a bumpy bend, and if anything breaks loose it will be the back end, with which most drivers can cope. These characteristics combine to make the TR3 a safe and satisfying car in the hands of both the novice and the expert. Its handling is nearly foolproof, in addition to being pleasantly fast.

The ride in general is quite good. The short wheelbase and shock settings produce a certain amount of pitching over big bumps, but smaller irregularities are nicely absorbed. On the whole, the suspension represents a fair balance between ride and roadholding. The abovementioned change in tire pressures will tip the scales either way, at your discretion.

Night driving finds the headlights sufficiently bright, but adjusted a little low on this car. Instrument lighting is very thorough, and a rheostat adjustment would be a welcome fitting. Driving vision with the hardtop is very good, both to the rear and forward over the broad, flat hood.

I had hoped for higher top speed figures, but adverse conditions and the low mileage of the test car prevented the attainment of the factory-claimed 110 mph maximum. All tests were run with two people aboard, and the recorded figures are thus the more remarkable. Similarly, consistent hard driving during the test prevented the establishment of a representative gas mileage figure, but I can practically guarantee that no TR3 owner will ever drive any less than 21 miles on a gallon of fuel. Overdrive top gear was a real luxury, since in that ratio the engine was turning over only 3,800 rpm when flat out at 100 mph.

The Triumph TR sports cars are designed to provide a maximum of performance at minimum cost, and this has been accomplished in an exhilarating manner. The TR3 represents the present peak of refinement of this type, and has much more to offer than the early cars. Several discordant elements remain, but they are not likely to perturb the enthusiastic driver, who will find that the TR3 will respond spiritedly to his every motoring whim. #

The Autocar, 28 September 1956

Disc Brakes for the TR3

ONE of the most significant trends in the post-war period has been the detailed development of successful basic designs from year to year. This is in contrast with the pre-war fashion when entirely new models were introduced each year, with seemingly complete disregard for the lessons learned from preceding designs. The two-litre Triumph TR3 is a good example of the modern trend, as it has been subjected to only detail improvements since its introduction at the Motor Show of 1952.

This two-litre sports car has built up an enviable reputation in International rallies and in sports car racing, its most outstanding successes being the sweeping victories in this year's Alpine Rally, when it won five Coupes des Alpes with six cars entered, and first five places in its class.

The major development for next year's cars is the fitting of Girling disc brakes at the front. With the enhanced performance achieved with the car, the present drum brakes cannot be further increased in size to achieve a higher level of braking efficiency. It will be recalled that the same disc brakes were used on the TR2 cars which competed successfully at Le Mans last year.

The discs are of the segmental pattern, with a fixed as distinct from a floating disc, and the pads remain close to the rubbing surface at all times. Thus displacement of the hydraulic fluid and idle travel of the brake pedal is kept to a minimum. The brake pad linings can be inspected at any time, and are easily renewed when worn out, by the removal of two screws and two triangular retaining plates. An additional advantage of this fitment is that the ratio of braking on the front has been increased to 60 per cent, at the same time achieving a longer wear life.

The Vanguard III type of rear axle with its Timken taper roller bearings is introduced, the difference in track necessitating special half-shafts. These modifications are introduced with no increase in price, and the TR3 continues to offer very high performance for a modest outlay.

The callipers of the disc brakes are stiff in section to resist deflection. There is an hydraulic operating cylinder to each pad, but only one bleed point is necessary. Clutch and brake master cylinders (right) are fed from a common reservoir, which has two concentric header chambers for this purpose

What do you look for in a sports car?

The New TRIUMPH TR-3 gives you the most for $2599

New from the word "Go!"—that's the exciting story of the new Triumph TR-3. New styling, new engineering, a new idea in fun-plus-convenience make this British-built beauty the sports car buy of the year.

The new Triumph TR-3's 100 horsepower, 1991 cc. engine gives you through-gears acceleration from 0 to 50 miles per hour in only 8 seconds. A few seconds later, you're flat-out at 110.

Zest-drive a Triumph soon. See how much more car you get for so little money

It's a **TRIUMPH**

$2599 plus tax and license at U.S. ports of entry. Wire wheels, hard-top and rear seat optional extra.

Parts and service are readily available from coast to coast.

STANDARD-TRIUMPH MOTOR COMPANY, Inc.
122 East 42nd Street, New York 17, N.Y.

For more information and dealer nearest you write: West of Mississippi — Cal Sales, Inc., 1957 West 144th Street, Gardena, Calif. • East of Mississippi — South Eastern Motors, Inc., 1937 Harrison St., Hollywood, Fla. • In Canada — The Standard Motor Co. *(Canada)* Limited, 496 Evans Avenue, Toronto 14

New Hard Top (optional) slips on the TR-3 faster than the weather can change.

New rear seat (optional) makes TR-3 your best fun and family sports car buy.

New Graceful chrome grille front-end accentuates the sports car look and action.

JUNE 1956

THE AUTOCAR, 27 JULY 1956

THE GALLAN[T]

The Autocar photographer P. Felkin hard at work amongst the mountain snows

BY any standards, the Triumph performance in the Alpine Rally was the outstanding one. Out of six finishers in the class, five were TR3s, all winning Alpine Cups, awarded for entirely clean sheets; the solitary Alfa Romeo which managed to intrude on this one-make fiesta was penalized 540 points. It would be nice to say that in trying a TR3 over the Alpine circuit *The Autocar* had anticipated this victory, but at least it can profit by the coincidence, in describing just why this all-British product was so good at high-speed mountaineering.

The departure of the TR3 from sports car tradition is very slight. The traditional in sports cars was a massive cruciform frame, half-elliptic springs front and rear, an engine of substantial size set well back in the frame, and overall ratios exactly suited to the intended job, which involved a high third gear so that acceleration could be prolonged to the last moment before the ultimate ratio was engaged. This, with the appeal of a very fast maximum speed in mind, was usually "about as high as it could stand." That was why (among other things) the engine of a classic sports car had to be of substantial size—in order to space out the ratios far enough.

The TR3 retains the cruciform-braced frame. It substitutes independent front suspension for the old beam axle, by coil and unequal wishbones, damped by telescopic struts at the front. At the rear piston-type dampers are retained with the half-elliptics. The engine is the well-known four-cylinder Standard Vanguard unit, with smaller-bore, wet cylinder liners substituted to bring it below two litres (1,991 c.c.). It is abaft the front cross member, which, itself, is about 10in back from the front of the frame. In TR3 form, it develops 117.5 lb ft maximum torque at 3,000 r.p.m. (145 lb per sq in b.m.e.p.) and goes on up to 5,000 r.p.m. to record maximum b.h.p. (100). A four-speed, centrally-shifted gear box is used with the Laycock-de Normanville optional overdrive, so that a ratio can be selected for almost any type of work, if the driver likes that amount of finesse; used as a cruising gear, the overdrive top of 3.03

Compact and powerful, the TR3 makes light of the highest road summits

MOUNTAINEER

to 1 is certainly in the traditional "as high as it can stand" category.

Now the great virtue of the classic sports car was the way it conveyed to the driver exactly what stresses the speed was imposing upon the road-holding, and the TR3 has inherited that ability. Independent suspension has removed real road shocks from the steering, but it remains sensitive to surface in a way that is nicely revealing. The judicious distribution of weight holds a correct balance between under- and oversteer, yet when, on a bend, there is any loss of adhesion between tyre and road, it is at the back, where correction is possible and easy. After a mile or two of driving, confidence comes that this car can be thrown about, and in over 3,000 miles of such throwing it showed never a hint of treachery.

An audible engine (again in the tradition) can be r.p.m.-judged precisely by the note from the wide-bore exhaust—always a help in the accurate balancing between traction and loss of it. High-speed changes are made more skilfully as a result, and if these occur at moments when the car tends to have only a tenuous hold of the surface, there is less risk of a substantial difference between engine revs and road speed being translated into a skid by a careless clutch engagement. As for the short, rigid gear lever, it is superbly in keeping, while the overdrive control is handy to the flick of a right-hand finger.

Here are the essentials of fast mountain driving, with corollaries in the high-geared cam and lever steering and the reliable brakes. At a weight of less than a ton, 175 sq in of brake lining will pull the Triumph up with consistency, until the screech of a rivet head tells you that it is time for new segments to do their share.

The seats keep the occupants well in place while all this exciting stuff is going on, and the one criticism which those occupants are likely to have is of the numerous excrescences—at least of the hard-top model—that manage to rap a funny-bone or a knee in the cockpit. These are the screen fixture sockets—it was amusing to note that Gatsonides had covered his four with sponge rubber pads —and the body side, which presents a painful angle to the thighs of the occupant who swings into his seat carelessly.

On only one type of surface were there any shortcomings in the design ensemble that provides the TR3 with its roadholding abilities; on a very much potholed surface there was sufficient wheel patter in traversing the road at speed for the car to slither a bit. Again, those who can recall driving truly traditional sports cars

Five thousand feet up in the Slovene Alps, at the top of the Vrsic Pass

of years ago will recognize the phenomenon, and may well, also, recall the pleasure of controlling a car that is more skating than rolling.

Enough speed is there for the most avid enthusiast: 100 m.p.h. was seen along the *autostrada* between Turin and Milan; the occasional straights out of the mountains would send the needle up above the 80 m.p.h. mark. So far as could be judged the speedometer was quite accurate, at least as high as 80 m.p.h. Between hairpins on the mountains, second and third, or second and overdrive second, could be used to make the most of the available space. The five clean sheets of the competitors are a better testimonial to the Triumph's abilities in this direction than the words of a camp-follower.

The five clean sheets, also, are a better testimonial to the engine's reliability—a virtue that is well known in Vanguard saloons. The Triumph team was running on Champion L11S plugs, a hot variety, while we were on the standard L10S; and we were burning plugs in the mountains. But here is an interesting point. We were also running on 100-octane AGIP fuel, being eager to try this fuel and to give the TR3 the best that money could buy. We burned a plug as we used our last tankful, then filled up with French Super and made for the finish of the rally. There we ran into Ken Richardson, the Triumph *chef d'équipe*, who immediately recommended the L11S plugs, so we bought five and started for home. The TR3 did not burn a single other plug on that run home along the flat main roads of France.

The question remains, therefore: Was it the mountaineering that cooked the plugs? Was it the 100-octane fuel? Was it the combination of the two? Personally I think it was the first, to the tune of at least 90 per cent, for there is not the slightest doubt that forceful mountain climbing in the heat of summer is a great strain on an engine. There is equally not the slightest doubt that the TR3 will withstand that strain. Don't take my word for it: take another look at the Alpine results.

M. B.

Job done—and this year's Alpine Rally press plates are removed from the TR3

This illustration shows the new positioning of the doors and the children's seat which is an optional fitting.

Triumph T.R. 3

FITTED WITH DISC BRAKES AT THE FRONT

THIS is the third time we have road tested a Triumph sports car since the original T.R. 2 was introduced in 1952. In five years, the Triumph has changed little outwardly, but all mechanical components have been continuously developed. The robust four cylinder push rod overhead valve engine is now more flexible, giving 100 b.h.p. at 5,000 r.p.m. as against the 90 b.h.p. at 4,800 r.p.m. on the earlier cars. A most important development is the fitting of disc brakes on the front wheels. This feature greatly enhances all round performance. The car can be driven very hard on twisting roads without being hampered and slowed by brake fade.

Regardless of mechanical developments, the most attractive feature of the T.R. 3 is its manner of going. In the medium speed ranges performance is entirely effortless. Even when the car is travelling at around 90 m.p.h. the engine never seems to be working too hard and although the general noise level increases rather sharply as road speeds rise, the power unit and transmission line components are obviously subjected to relatively low stresses as the machine approaches terminal velocity. It is this effortless performance which, almost as a by-product, allows such remarkable fuel economy when cruising at speeds up to about 85 m.p.h. We can hardly forget the result of the contest which took place some time ago, between a T.R.3 car and an Auster aircraft. The Triumph not only used less fuel in making a journey from the south of England to John O'Groats, but also bettered the Auster's time for the same journey. Mr Alick Dick, managing director of the Triumph company made an appropriate comment at that time—"if you have time to spare, go by air."

As when driving most sports cars, there is some tendency to make overmuch use of the gearbox on the Triumph. Even on the open road one tends to slip smartly into third gear when overtaking. The short central control lever does invite use, of course, but there is really no need to make such gearchanges. The power to weight ratio of the car is good, and the engine is flexible enough to pull heartily from fairly low speeds. Heavy traffic can be negotiated at 30 m.p.h. in top gear, without much difficulty, and the car can be started in second gear with ease. At the same time, it is obvious that the engine produces its real power at speeds above 2,000 r.p.m. From this point the car surges forward and the driver must keep an eye on the revolution counter if the recommended 5,000 r.p.m. limit is to be observed. A curious feature, common to all the Triumphs we have ever driven, is the exhaust resonance that occurs at about 2,200 to 2,400 r.p.m. It could be that the exhaust pipe length is tuned to increase performance at relatively low r.p.m. but it might be better if the resultant blare could be avoided since the car is apt

The latest version of the now famous T.R.3 engine is fitted with a matched pair of large S.U. carburetters. The inlet and exhaust arrangements on this engine are particularly interesting

The engine suffers no lack of accessibility when installed in the chassis. The carburetters, distributor, dipstick and oil filler cap are easy to reach

to attract unwelcome attention when motoring at quite modest speeds in heavily populated areas.

OUR test car was not fitted with the Laycock de Normanville overdrive unit which is offered as an optional extra on the T.R. 3. This overdrive unit does, without a doubt, bring about a slight increase in performance, especially between 55 and 90 m.p.h., and also adds some 3 to 4 m.p.h. to the maximum speed of the car. The overdrive unit also helps to keep engine revolutions down, although it should be mentioned that the limit of 5,000 r.p.m. recommended by the makers can be exceeded, briefly, without loss of reliability. We know of one much rallied T.R.3 which has been taken to 6,000 r.p.m. in the intermediate gears on many occasions without trouble. As was mentioned earlier, mechanical noise increases as engine speeds rise. This is most noticeable when the car is used in its closed form. The noise in itself, is not unpleasant and can be traced to the push rod overhead valve gear. The large tappet clearances necessary on the Triumph make the engine noisier when cold, a slight improvement takes place as normal running temperature is attained.

One apparently minor modification to the bodywork on the T.R. 3 greatly improves the utilitarian value of the car. On the early cars the doors extended below floor level, so that they were apt to foul pavements and etc., when opened. On the latest models the floor is extended to form a sill so that the bottoms of the doors are now some four inches higher. The door catches, and the bonnet and boot door locks, on the Triumph are rather unusual and could possibly be improved. The doors have no external handles and can only be opened by reaching through the sidescreen flap. The bonnet top and luggage compartment lid can only be opened by making use of a "coach key," and since the luggage compartment can also be locked with the same key that is used for the glove locker on the facia this means that one must undo three different catches to remove or insert luggage. The quick release petrol filler cap, which is located centrally above the luggage compartment, is easily reached from either side of the car. With the hood lowered no exhaust fumes are sucked back into the cockpit, but careless pump attendants can spill fuel inside the car.

THE simple body lines and general outline of the T. R. 3 are purposeful and functional. The car is easily kept clean. All panel work is smooth and there are no indentations which are difficult to wash and polish. Internally, the car is remarkably roomy. With the hood up the car is not difficult to enter and, once seated, front seat passengers are generously accommodated. The children's seat, which can be optionally installed at the rear of the car is simply intended for the youngsters, but can be very useful when carrying parcels and loose pieces of luggage inside the car. The driving seat has a good range of adjustment and the controls are very nicely positioned, although we were surprised to find that "heel and toe" changes could not be made on the car tested. This may have been due to maladjustment of the throttle pedal since we recall having made such gearchanges in other Triumphs. The central gear lever mounted on the prop-shaft tunnel is in an ideal position. The fly-off handbrake is also convenient in use and is powerful in action. The dashboard carries a full complement of instruments. Directly in front of the driver, on either side of the steering column, there is a matched speedometer and revolution counter. When the car is running in top gear the speedometer and rev. counter needles move in the same plane, and in unison. A small central instrument panel accommodates four matching dials which comprise of an ammeter, fuel gauge, thermometer and oil pressure gauge. The ignition switch, choke and starter motor controls can be operated with one hand. An effective re-circulating type heater was fitted on the test car.

Smooth in appearance, the T.R.3 body is functional and yet of pleasing aspect. Several minor improvements to bodywork and fittings have been made since the car was originally introduced

As far as the interior layout of the car is concerned, there is little to criticise. We are personally prejudiced against the facia top mounted central style rear view mirror but this is somewhat unavoidable in an open car. The hood itself is an excellent example of its type. The plastics coated fabric covering can be completely removed from the folding frame and is easily stowed. With the top up the car is capable of recording a slightly higher speed than when it is used in open form. The hood shows no desire to flap at high speed and the latest type Perspex panelled side screens are a great improvement, but produce periodic buzzing vibrations. The layout beneath the bonnet also deserves praise. The large twin S.U. carburetters ignition distributor, oil filler cap and battery etc., are easily worked on. We were particularly impressed with the neat hydraulic brake and clutch operating cylinders, which are very accessible and can be checked in a moment.

THE new brake layout, which incorporates Girling disc brakes on the front wheels, is of real benefit when the car is being driven hard. At relatively low speeds no great improvement, as compared to the more normal previous arrangement, is at once noticeable. The pedal pressure required is neither high nor low, but firm pressure brings nicely graduated results. On roads such as the Loch Lomondside and Arrochar-Helensburgh routes the car can be driven very hard, braking heavily before each of the numerous sharp bends, without signs of fade or erratic operation. Those sporting drivers who have, no doubt, driven "normal" cars hard over these roads during night rallies will appreciate the full meaning of this statement. Most low cost cars would suffer fairly drastic fade in the hands of a hurrying rally driver on the stretch between Helensburgh and Arrochar, and we mustn't forget that the Triumph is, after all a low cost car. The basic price of £680 is in fact remarkably low. There can be no doubt that the T.R. 3 is one of the cheapest high performance machines available anywhere in the world. The performance characteristics of the car make it suitable for everyday use and the interior accommodation is highly satisfactory from a touring point of view. For long fast journeys, with two persons and a fair amount of luggage aboard, the Triumph offers performance and economy that few cars can approach. With more than 100 m.p.h. available at the end of a relatively short straight, and the ability to cruise continuously at speeds between 80 and 90 m.p.h. where conditions allow, the Triumph can most certainly be regarded as a desirable machine. The safe handling characteristics of the car, coupled with a really dependable braking system, favourably accentuate these points.

MECHANICAL SPECIFICATION

Engine

Cylinders—4
Bore—83 m.m.
Stroke—92 mm.
Cubic Capacity—1,991 c.c.
Valves—Push rod, overhead
Compression ratio—8.5 to 1
Max. Power—100 b.h.p.
    at   5,000 r.p.m.
Carburetters—Two semi-downdraft S.U. type H.6

Transmission

Top gear (s/m)—3.7 to 1
3rd gear (s/m)—4.9 to 1
2nd gear (s/m)—7.4 to 1
1st gear 12.5 to 1
M.P.H. per 1,000 r.p.m. in top gear—20.2

Chassis

Brakes—Disc pattern on front wheels. Drum type at rear, Girling hydraulic operation.
Suspension: front—coil spring and wishbone i.f.s.
    rear—semi-elliptic, leaf springs
Dampers—telescopic hydraulic at front, piston type at rear
Wheels—bolt on steel disc (wire type optional)
Tyres—5.50 x 15

Steering

Steering gear—cam and lever mechanism
Turning circle—32 feet
Turns of wheel, lock to lock—2¼

Dimensions

Wheelbase—7 feet 4 inches
Track: front—3 feet 9 inches
    rear—3 feet 9½ inches
Overall length—12 feet 5 inches
    width—4 feet 7½ inches
    height—4 feet 2 inches
Kerb weight—19 cwts.

PERFORMANCE FIGURES

Weather—Fine, warm with light breezes. Temp. approx. 60 deg. F.

Fuel—Premium Grade.

Speedometer Correction—

Reading	20	30	40	50	60	70	80	90	100
Timed speed	18	28	37	47	58	68	78	88	98

Maximum Speeds—

Gear	M.P.H. (Normal and Maximum)
Top	Mean 103.6 Best 104.2
3rd	76 (5,000 r.p.m.)
2nd	30/50
1st	15/28

Acceleration (Time in seconds)

M.P.H.	Top	3rd	2nd	1st
10-30	9.9	7.0	5.5	—
20-40	9.0	6.0	5.8	—
30-50	9.2	6.1	—	—
40-60	9.4	6.6	—	—
50-70	9.9	7.0	—	—
60-80	11.0	—	—	—

From rest through the gears—

0-30	3.9 secs.	0-70	14.3 secs.
0-40	5.7 ,,	0-80	20.0 ,,
0-50	7.7 ,,	0-90	27.3 ,,
0-60	11.0 ,,		

Fuel Consumption—

Driven Hard—28.2 m.p.g.
Normal—35/37 m.p.g.
Fuel tank capacity—12½ gallons

Price—Basic. £680; Purchase Tax, £341 7s 0d; Total, £1,021 7s 0d.

(The Triumph TR3 used during this test was kindly placed at our disposal by Messrs. McHarg, Rennie and Lindsay, Berkeley Street, Glasgow.)

DEBUT. No excesses and no success either, in SYA's first outing at a club speed trial.

A TALE of TWO-SEATERS

By RICHARD BENSTED-SMITH

Repeat Order for a Maid-of-all-Work; the Triumph TR2

THE one-car motorist and the one-make man are by no means related. Restricted by economic circumstance to the possession of one car I cannot really afford—a situation shared by the great majority of private owners—I am generally of too impatient a disposition to owe allegiance to any make of car for more than a year or so. Which makes it surprising that, after a brief lapse into more sociable but less satisfying machinery, I recently did some quick sums and bought another TR2.

On the face of things, the Triumph might not appear the most intelligent choice for a one-car type. Living in London, W.2, commuting to E.C.1, and normally covering about 100 business and pleasure miles a week within the Postal Area is an unusual life for a 2-litre sports car. And what, you might say, about economy, and what about the price, and what about having only two seats in your only car?

Working through the questions backwards, I share with a number of business motorists the fact that nine-tenths of my driving is done with only one man in the car, let alone four. Furthermore, without being actually a recluse by nature I must admit that the ability to take only one passenger can at times be a tactful convenience. At least one can be sure that anybody accepting a lift in the back of the Triumph is either a good friend or a real hard-luck case. As to the price, I could not afford it, but that is neither here nor there, as the same was true of other cars much less suitable to my needs, while more expensive cars meant finding money that could not even be borrowed. The problem after all is to find something you can contrive to buy, can afford to run, and like driving. The TR2 is (apart from insurance) a comparatively inexpensive car to run, provided you do not break it; then, like any specialist car which packs half a gallon into a two-pint pot, it takes a certain amount of fairly expensive unpacking.

A word of explanation becomes necessary at this point. Seven out of the eight cars which have passed through my hands in as many years (impatient, did I say?) have been acquired secondhand, and the Triumphs were not exceptional. Given a certain degree of wariness, mechanical aptitude, and sales resistance, it is possible to buy even a high-performance car when somebody else has finished with it. *Caveat emptor* indeed; but it does not follow that all vendors are dishonest, nor that they wish to sell because their car is no longer of any value.

Triumph number one entered my life towards the end of 1954, when the type had been in existence just long enough to prove its ability to win almost any kind of competition hands down. SYA 176 was six months old at 7,000 miles, which implied that it had not been altogether inactive. In fact, paintwork in two shades of off-

A TALE of TWO-SEATERS

FRUGALITY proved a most rewarding virtue in 1955.

ALL-ROUND distinction brought home the Vintage S.C.C.'s all-age handicap award, the Pomeroy Trophy. By means of mathematics cars like the 1914 Grand Prix Vauxhall (suitably equipped with wings and things) can compete on level terms with the younger generation.

white disclosed the fact (of which there was no secret) that the first owner had rather inconsiderately driven it through a hedge. As the Standard Motor Company readily confirmed, a new chassis frame had been included in the subsequent rebuild.

In the following fourteen months the car covered another 14,000 miles. Being a journalist's transport it forcibly adopted journalists' habits, which are well known to be erratic, with alternate spells of over- and underworking and generally irregular feeding. It stood it remarkably well. I think that if one can pin down a definition of the much-used expression of the moment, Gran Turismo, or Grand Touring, it should be to describe a car in which long distances can be covered not only fast, but comfortably and *habitually*, without wear on man or machine. Examination of a record of weekly mileages for 1955 shows that between February and September there were five occasions when the total for two consecutive weeks was more than 1,000 miles—in one case 1,200 miles. Now I am aware that for many people 500 miles a week is not out of the way. These records, however, have been taken from Sunday morning to Saturday night, so that the exceptional totals generally involve a concentrated working week-end. On such occasions as Easter in the West Country, where Trengwainton hill-climb follows the Land's End Trial, and Friday, Saturday, Sunday, Monday and Tuesday run up 938 miles, I have been glad of the Grand Touring Triumph.

Negligible Driver Wear

In the modern manner, it requires greasing at only 1,000-mile intervals, so that if, as seldom occurs, one remembers to attend to the chassis regularly, such long journeys are taken care of; if one does not, it appears to make little difference. With such maintenance, or lack of it, the TR2 would show a fairly consistent fuel consumption of 31 m.p.g. (no overdrive was fitted) for the miles away from London, use oil at about a pint in 600 miles, and no water to speak of. Driver wear was negligible, a fact which seems to me directly attributable to the successful formula for a classic touring car: take a light, compact chassis, and install in it a large, unstressed engine of basically "cooking" design. There was a tendency amongst some vintage enthusiasts, which has subsided in the shadow of a growing list of honours, to decry the TR2 as not a "real" sports car; yet the 30/98 Vauxhall of famous memory applied the same formula, and left a reputation as a touring car which often outshone the sports cars of its day.

To return to SYA, however; it was on arrival in completely normal, well-worn trim, apart from the piebald paintwork already mentioned. Oil pressure stood at about 50 lb./sq. in. at 2,000 r.p.m., which is 20 lb. less than the service manuals say, but appears to be a figure these engines settle down to after some running. It remained constant for over a year, when a combination of circumstances caused the demise of a big-end. One pair of tyres was little worn (I suspect they had already been replaced) whilst the others had only about half the tread remaining. Like all cars put up for sale it had the better tyres on the front, and like a minority of drivers I kept them there, preferring wheelspin to loss of steering on a wet road. It may be because of the vicious circle of worn tyres and spinning wheels that with Triumphs I have found a tendency for those on the back wheels to wear faster. Judging by other people's experiences, however, it is more likely to be because I would rather brake

TOURIST without a trophy. The TR2 may well have held the post-war record on Donington circuit. A reconnaissance in the spring of 1955 revealed very few competitors.

hard before a corner than wait and see if I have to brake harder still half-way round it. The latter method probably takes you to your destination a little sooner, besides saving the brakes some of the time; but only after a further 9,000 miles were the worn back tyres replaced, and 5,000 miles later there was still no need to do anything about the front ones.

Care and Maintenance

Brakes, on the other hand, do comparatively badly on this kind of driving, although with a fair proportion of urban mileage included they never became due for relining through wear. That first TR2, however, came of the early series, which suffered from a tendency to grabbing when the brakes were cold, and the higher proportion of braking effort on the front wheels could lead to fade if one took to hurrying down the hills of north Devon. A cure for both complaints was achieved by relining the front brakes only on the recommendation of Ferodo, with a different material, after which they behaved perfectly on all occasions. A minor irritant on the first Triumph was a leaking brake reservoir cap, which spilled paint-solvent fluid and proved surprisingly difficult to replace through the spares people. Probably on account of dirt or age both cars have been rather prone to acquire air in the hydraulic system, not dangerous but a tiresome job to put right and one that is impossible without assistance.

Self-service is a habit I must admit to having dropped with the Triumphs. A believer in jacking up the body to take the weight off front suspension parts for greasing, especially when no pressure-lubricating equipment is available, I rebel at the use of one of the worst screw-jacks fitted to any modern car. Battery and dip-stick are easy to reach and the car benefits accordingly. I have fitted a new water thermostat with little trouble, and a new fan belt with a lot—I would probably do it myself again if necessary, simply because the time it takes would bring a disproportionate labour charge for what ought to be a

TWO-UPMANSHIP. The second TR2 boasts refinements like wire wheels and doors which can be opened above the kerb.

THE MOTOR

A TALE of TWO-SEATERS

TOWN OR COUNTRY brings out the best in a car which will potter through Hyde Park at 20 m.p.h. or Salisbury Plain at 100 m.p.h., in the same gear.

simple operation. Beyond that, I would leave it to professionals with specialized know-how and tools, which is only reasonable with a specialized car. For such a machine it is very free of small troubles. The most absurd was when a gear lever stuck in top gear on account of a pebble in the uncovered remote-control selectors, and the car was then driven 20 miles home across London (by night) using only the unprotesting clutch.

This is a particular tribute, and bears out my classic-car formula, for direct top gear works out at 20 m.p.h. per 1,000 r.p.m. The earlier allusion to a half-gallon in a two-pint pot was not made by chance. Two litres of Triumph engine propel a chassis based upon the one-litre Standard Eight, so that much less than the maximum 5,000 r.p.m. will take you to most places with ease and comfort. Even with the major modifications to frame and suspension adopted for the Triumph one could hardly expect this two-dimensional chassis to be a prodigy of stiffness and roadholding, which makes it the more valuable as well as entertaining to be able to go very fast between corners.

A brief and unsuccessful skirmish in a club speed-trial opened a brief and otherwise wholly successful competition season for SYA. That we were no better than seventh in the 2½-litre class may be chiefly ascribed to the fact that six other drivers (four of them in TR2s) drove faster. An interesting and less direct comparison with other cars occurred shortly afterwards when the Triumph, followed by a sister model, beat the complex handicap in the Vintage Sports Car Club's Pomeroy Trophy competition. As a potted all-round test of cornering, braking, acceleration, speed, endurance and economy the Pomeroy Trophy is unique, by virtue of a handicap which, in theory, equalizes cars of every age and size from 0 years and 1,950 c.c. upwards. The whole thing is fairly light-hearted, and with preparation amounting to checking the oil level and tyre pressures the TR2 galloped cheerfully up and down and round Silverstone, finally clinching things on a fuel consumption of 30 m.p.g.

Economy Brakes

Ambition thereupon reared its head. With the appearance of the regulations for the first British Mobilgas Economy Run, which left practically no holds barred in either tuning or driving methods, the services of the Standard Motor Co. were enlisted. The Engineering Department in Coventry advised, and the Service Department in London operated to the extent of grinding valves, fitting a new thermostat and tuning the S.U. carburetters weak. We blew the tyres up, used a lot of luck and gravity for 600 miles, and came back with a figure of 71 m.p.g. As with the Vintage club's event, the rules for the Economy Run were changed in the following year, to the detriment of two-seaters, which is perhaps fair enough as a practical measure. Looked at more objectively, it does measure the general excellence of the design as personal, if not family, transport. The Triumph is an inherently economical car, but two factors contributed to this extreme petrol-squeezing; an engine which would pull it up hills in top gear when smaller-capacity cars had to change down and use more revs; and sports-car brakes on which one could rely during prolonged coasting. Looking back at the account of our journey, I see that the descent of the Long Mynd involved a 1,000-foot change of altitude, with a control near the bottom. Where we coasted all the way with a dead engine, at least one family saloon finished the descent in bottom gear, and then overshot the halt with faded brakes.

Near Perfection

The first Triumph was sold, and eight months later replaced by another. If it suits your pocket and your normal burden of passengers and luggage, so useful an all-rounder seems too good to be without. The "new" TR2 is an advance on the old one, for it has the later, higher door sills, wire wheels and much better visibility through the rear window. It has Michelin X tyres which vastly improve cornering power in the wet, although they impose a penalty in heavier steering. It has, to my mind, a more awkward (but more foolproof) bonnet release by carriage-key instead of remote control.

It shares with the earlier model two comfortable seats and more than enough length for my long-legged six foot shape, space behind the seats, a lockable boot which really holds things, and a lockable cubby which makes all the difference to an open car. Around town it slips through gaps in the traffic that even belligerent taxi-drivers do not bother about, and on the open road it has that comforting feeling that whether you want to go faster or slower there is always something in reserve. It is not perfect, but it comes as near perfection as you can expect from a mechanical maid-of-all-work.

Neither rain, wind nor etcetera can stay this courier

TRIUMPH TR3
only $2675*

For unfailing delivery of the most exciting performance, get behind the wheel of a TR3. Soft top up, against early spring wind... seated deep in cushioned leather... you touch the starter and you're gone!

What obedience... you have a tiger in your power. Gently purring through traffic, ready to roar out on the open road, you're feeling (and causing) a new sensation. Effortlessly you have her up to 80 in Overdrive. A quick down-shift and you're taking the horseshoe bend without a sway, leaving the "big jets" braking and shaking behind. And as you search out a lonely road to try her straightaway power, it's nice to know that your disc brakes† have stopping power to match.

Yes, this is British engineering brilliance... the world's finest performance value. And the TR3 delivers this in the greatest of style. You'd be wise to arrange for delivery now. Guest-drive the TR3 today.

*2675 at U.S. ports of entry, plus state and/or city taxes (slightly higher West Coast.) Wire wheels, hard top, rear seat, white wall tires and overdrive, etc. optional extra
SPECIFICATIONS:
BRAKES: *Disc brakes on front wheels†*
TOP SPEED: *110 MPH* **MILEAGE:** *up to 35 MPG*
ENGINE: *4 cyl. (OHV) 1991 cc* **OUTPUT:** *100 BHP*
ACCELERATION: *0-50 in 8 sec.*
MAINTENANCE: *Parts and service available coast to coast! Free Brochure and dealer list on request. Write now—for fun!*
†*A Triumph-plus... as standard equipment.*

STANDARD-TRIUMPH MOTOR COMPANY, INC., Dept. J5, 1745 Broadway (at 56th St.), New York 19

The Motor Road Test No. 26/57 (Continental)

Make: Triumph **Type:** TR3 Hardtop (with overdrive)
Makers: Standard Motor Co. Ltd., Banner Lane, Coventry

Test Data

World copyright reserved; no unauthorized reproduction in whole or in part.

Conditions: Weather: Mainly dry, light crosswind. (Temperature 54°-63° F., barometer 29.5-30.1 in. Hg.) Surface: Smooth concrete (Ostend-Ghent motor road). "Road Speed" tyres at 40 lb./sq. in. as advised for fast driving. Fuel: Premium grade Belgian pump fuel (92 RM Octane).

INSTRUMENTS:
Speedometer at 30 m.p.h. ... 7% fast
Speedometer at 60 m.p.h. ... 7% fast
Speedometer at 90 m.p.h. ... 8% fast
Distance recorder ... accurate

WEIGHT
Kerb weight (unladen, but with oil, coolant and fuel for approx. 50 miles) ... 19¾ cwt.
Front/rear distribution of kerb weight ... 53/47
Weight laden as tested ... 23¼ cwt.

MAXIMUM SPEEDS
Flying Quarter Mile
Mean of four opposite runs ... 109.1 m.p.h.
Best one-way time equals ... 110.8 m.p.h.
"Maximile" Speed (Timed quarter mile after one mile accelerating from rest)
Mean of four opposite runs ... 106.2 m.p.h.
Best one-way time equals ... 107.1 m.p.h.
Speed in Gears (at 5,000 r.p.m. recommended rev. limit)
Max. speed in direct top gear ... 100 m.p.h.
Max. speed in O/D 3rd gear ... 93 m.p.h.
Max. speed in direct 3rd gear ... 76 m.p.h.
Max. speed in O/D 2nd gear ... 61 m.p.h.
Max. speed in direct 2nd gear ... 50 m.p.h.
Max. speed in 1st gear ... 31 m.p.h.

FUEL CONSUMPTION
(Overdrive top gear)
41½ m.p.g. at constant 30 m.p.h. on level.
43 m.p.g. at constant 40 m.p.h. on level.
42 m.p.g. at constant 50 m.p.h. on level.
39 m.p.g. at constant 60 m.p.h. on level.
35½ m.p.g. at constant 70 m.p.h. on level.
29 m.p.g. at constant 80 m.p.h. on level.
26 m.p.g. at constant 90 m.p.h. on level.
(Direct top gear)
39½ m.p.g. at constant 30 m.p.h. on level.
41 m.p.g. at constant 40 m.p.h. on level.
38 m.p.g. at constant 50 m.p.h. on level.
Overall Fuel Consumption for 1,737 miles, 66.3 gallons, equals 26.2 m.p.g. (10.8 litres/100 km.).
Touring Fuel Consumption (m.p.g. at steady speed midway between 30 m.p.h. and maximum, less 5% allowance for acceleration), 34.0 m.p.g.
Fuel Tank Capacity (maker's figure), 12 gallons.

HILL CLIMBING at sustained steady speeds
Max. gradient on overdrive top gear: 1 in 11.7 (Tapley 190 lb./ton).
Max. gradient on direct top gear: 1 in 9.3 (Tapley 240 lb./ton).
Max. gradient on overdrive 3rd gear: 1 in 8.4 (Tapley 265 lb./ton).
Max. gradient on direct 3rd gear: 1 in 6.6 (Tapley 335 lb./ton).
Max. gradient on overdrive 2nd gear: 1 in 5.0 (Tapley 435 lb./ton).
Max. gradient on direct 2nd gear: 1 in 4.1 (Tapley 535 lb./ton).

STEERING
Turning circle between kerbs: Left, 33 ft.; right, 31¾ ft.
Turns of steering wheel from lock to lock: 2¼.

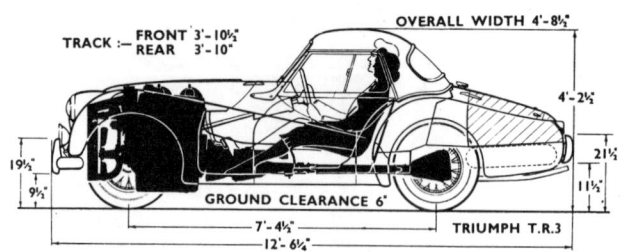

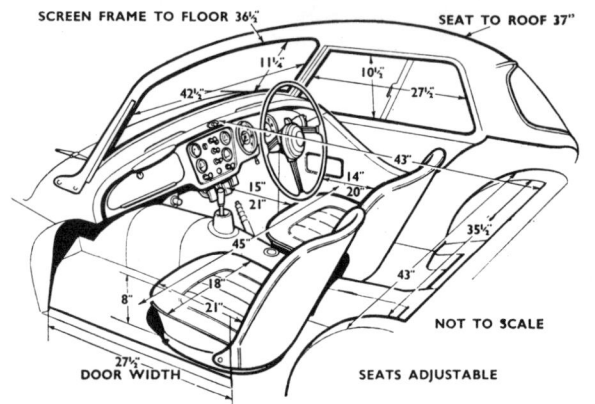

ACCELERATION TIMES from standstill
0-30 m.p.h. ... 4.6 sec.
0-40 m.p.h. ... 6.8 sec.
0-50 m.p.h. ... 8.9 sec.
0-60 m.p.h. ... 11.4 sec.
0-70 m.p.h. ... 15.6 sec.
0-80 m.p.h. ... 20.7 sec.
0-90 m.p.h. ... 28.2 sec.
0-100 m.p.h. ... 39.5 sec.
Standing quarter mile ... 18.5 sec.

ACCELERATION TIMES on upper ratios

M.P.H.	Overdrive top gear	Direct top gear	Overdrive 3rd gear	Direct 3rd gear
10-30	—	10.2 sec.	9.1 sec.	6.9 sec.
20-40	12.2 sec.	9.4 sec.	8.7 sec.	6.9 sec.
30-50	12.0 sec.	9.4 sec.	8.3 sec.	6.4 sec.
40-60	13.0 sec.	9.5 sec.	8.4 sec.	6.7 sec.
50-70	13.8 sec.	10.0 sec.	7.8 sec.	6.9 sec.
60-80	15.6 sec.	11.3 sec.	10.2 sec.	—
70-90	21.0 sec.	14.4 sec.	13.9 sec.	—
80-100	26.6 sec.	18.0 sec.	—	—

BRAKES from 30 m.p.h.
0.99g retardation (equivalent to 30½ ft. stopping distance) with 120 lb. pedal pressure.
0.92g retardation (equivalent to 33 ft. stopping distance) with 100 lb. pedal pressure.
0.81g retardation (equivalent to 37 ft. stopping distance) with 75 lb. pedal pressure.
0.53g retardation (equivalent to 57 ft. stopping distance) with 50 lb. pedal pressure.
0.30g retardation (equivalent to 101 ft. stopping distance) with 25 lb. pedal pressure.

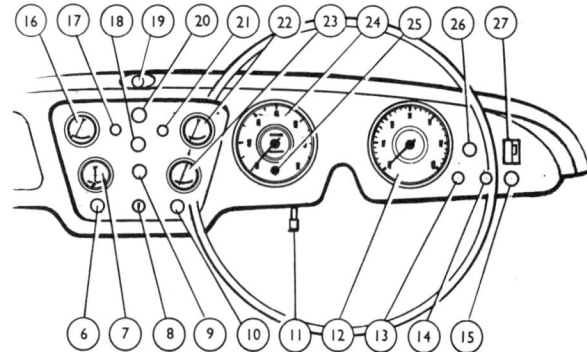

1, Headlamp dip switch. 2, Gear lever. 3, Handbrake. 4, Direction indicator switch. 5, Horn. 6, Starter switch. 7, Ammeter. 8, Ignition switch. 9, Lights switch (side, head). 10, Choke control. 11, Trip recorder adjusting knob. 12, Revolution counter. 13, Windscreen washer control. 14, Fog lamp switch. 15, Long-range lamp switch. 16, Fuel contents gauge. 17, Dynamo charge warning light. 18, Windscreen wipers switch. 19, Scuttle vent lever. 20, Panel light switch. 21, Direction indicator warning light. 22, Oil pressure gauge. 23, Water thermometer. 24, Speedometer with distance recorder. 25, Headlamp main beam warning light. 26, Heater fan switch. 27, Overdrive switch.

The TRIUMPH TR3 Hardtop

(with overdrive)

Even More Performance at Moderate Price from the Well-known Two-seater

THE character of the Triumph TR models has become so well known, in the three years since production examples first appeared, that it would be easy to imagine that no significant changes had accompanied small "face-lifting" operations upon the visible exterior. It is true that the TR3 is still essentially a fast, economical and roadworthy two-seater, built for fun but not for amusement only, with unusually good provision for luggage. Nevertheless there have been improvements.

In 1954 the 90 b.h.p. TR2 stood in the front rank of sports cars offering high performance at moderate price. Today an advertised 100 b.h.p. endows it with considerably increased maximum speed and acceleration which once again place it extremely well on a rating of speed-for-price. It may be a sign of the times that in the past year emphasis has clearly been placed on performance at the top end of the scale. The third and latest TR to be tested by *The Motor* recorded a best one-way maximum speed of very nearly 111 m.p.h. and acceleration from rest to 90 m.p.h. in the creditable time of 28.2 seconds.

More important still, however, the accent is put on really high speed as an everyday occurrence by the fitting of Girling disc brakes to the front wheels. The use of discs on a car with a basic price of less than £700 is final proof that this method of stopping has "arrived" for the ordinary motorist, and their effectiveness in day-to-day use is of great interest. In this case the standard braking performance figures given by Tapley meter tell only part of the story. The rest may be summed up by saying that in 1,500 varied but almost all hard-driving miles only very heavy rain appeared to have any effect upon the brakes, and that a temporary one. For practical purposes it was impossible to drive the car so hard on a public road as to reduce their effectiveness by overheating. To confirm this impression, an improvised test was made upon a motor road. After determining the pedal pressure needed just to lock the wheels from 30 m.p.h. under normal conditions, the Triumph was brought to rest from 100 m.p.h. three times in quick succession at a mean deceleration of 0.65g—that is with about the maximum effort usable from such speed in consideration of the presence of other traffic.

After three stops the pressure required to lock the wheels was unaltered, although the discs were hot enough to ignite a match held against their surface.

The fastest cars are constrained at times, however, to amble at walking pace through city traffic, conditions under which brakes meant for hard use are not always at their best. It is satisfactory to report a complete lack of temperament in this respect from the Girling units. The only time their performance was upset was on cambered and streaming wet Continental roads, where after a time the right-hand disc began to take less than its fair share of the load, causing the car to pull to the left. Braking was restored to normal as soon as the road became dry.

Having begun with the brakes, it is logical to pursue a report on the Triumph by reference to the other qualities tending to roadworthiness. No pretence could be made that the chassis and suspension are of advanced design, and the result of orthodox layout is the expected compromise between comfort and road-holding. Comfort is quite adequate; road-holding is, of course, much in advance of the average saloon, although short of the standards set by more expensive sports cars. Fairly

STOPPING power. The Girling disc brakes fitted to the Triumph's front wheels proved virtually impossible to fade. Only very wet roads had any marked effect on their performance.

In Brief

Price (including overdrive as tested): £787 10s. plus purchase tax £380 2s. equals £1,137 12s.
Price without overdrive (including purchase tax), £1,073 17s.
Capacity ... ... 1,991 c.c.
Unladen kerb weight ... 19¾ cwt.
Acceleration:
  20-40 m.p.h. in direct top gear ... ... 9.4 sec.
  0-50 m.p.h. through gears 8.9 sec.
Maximum direct top gear gradient ... ... 1 in 9.3
Maximum speed ... ...109.1 m.p.h.
"Maximile" speed ... ...106.2 m.p.h.
Touring fuel consumption ... 34.0 m.p.g.
Gearing: 20.2 m.p.h. in top gear at 1,000 r.p.m. (overdrive, 24.6 m.p.h.); 33.4 m.p.h. at 1,000 ft./min. piston speed (overdrive, 40.8 m.p.h.).

The TRIUMPH TR3 Hardtop

BUCKET seats, well raked, more than ample leg-room and controls in the right places stamp the TR3 as a serviceable sports car.

LUGGAGE capacity is exceptional for this type of car, and includes a lockable boot and facia glove box. The spare wheel is carried in a separate tray beneath.

stiffly sprung and with a low centre of gravity, the TR3 can, like any high-performance car, be made to slide under certain conditions, but what those conditions are, and how it reacts to them, depend to a greater extent than usual on the type of tyres fitted and their inflation pressure.

In deference to the speeds which would be maintained during performance testing on the Belgian motor road, the car came into our hands with the optionally available Dunlop Road Speed tyres, and a request from the Dunlop engineers that a pressure of 40 lb./sq. in. should be used for testing, whereas for everyday driving a pressure of, at the most, 30 lb./sq. in. proved necessary for acceptable riding comfort. On dry roads the Road Speed tyres showed good cornering power and freedom from squeal, but in the wet they left a good deal to be desired, the back wheels sliding on slippery corners at the smallest touch of throttle. Upon its return to England, the Triumph reverted to the normal Dunlop covers supplied as standard equipment, and the pressure was adjusted to 26 lb./sq. in. The result, confirmed by experience with other TR models, was noticeably better roadholding at the speeds likely to be obtained on most occasions in this country, accompanied by some noise from the tyres in sharp corners, and slightly heavier steering.

The two last are further accentuated by adoption of the recommended pressures for "normal" driving of 22 lb. and 24 lb. respectively. Whether or not a differential between front and rear pressures is maintained is really a matter of personal taste. Softer tyres on the front wheels serve to increase the understeering tendency which is inherent but not objectionable. The steering itself is again a compromise, free from backlash and sponginess, reasonably high-geared at two-and-a-third turns between locks, but without quite the instant response preferred by many sports car drivers. The steering wheel is large and well placed, and the Triumph has a turning circle as compact as that of most small family saloons.

Practical

Details of this sort are typical of a sports car which could almost be summed up in one word: practical. It would be hard to find a cockpit and driving position better thought out to suit most owners for most of the time. The seats are comfortable and well "bucketed," with backs fairly sharply inclined and on the low side, but the seats will slide far enough to accommodate the longest or the shortest driver. Allowance being made for their low height above the kerb, the seats are easy to enter. Instruments are large, legible and placed in exactly the right order to present their information with the minimum of strain. The hanging pedals are well placed, with space for the left foot below the clutch pedal for long uninterrupted periods, or with it slightly drawn back resting on the dip-switch. In this position the left leg leans against the handbrake, a fault easily forgiven for the convenience and excellence of the brake and its fly-off lever.

Remote-control central gear levers are all too rare nowadays. That on the Triumph is good by any standards, and the easy change in combination with Laycock overdrive on the upper three gears provides so many ratios, with so little effort, that it would be quite difficult to find oneself caught in the wrong gear. In fact, as is usually the case with this arrangement, some pairs of ratios such as overdrive third and direct top are very close, so that they are almost interchangeable according to which can be most quickly selected.

In spite of the number of gears available, their use can be largely cut out in leisurely driving, for the two-litre engine is exceptionally flexible and the car can quite easily potter through traffic at 20 m.p.h. in top. Detail modifications in the induction system seem to have spoiled the first-touch starting of earlier cars of this type, but very little use of the choke is subsequently needed. A vibration period causes slightly

HARDTOP adds a total of £52 10s. to the price of the basic model. Wire wheels figure in the list of a score of optional extras.

NEAT and distinctive in appearance, the Triumph undoubtedly owes both performance and economy to its clean lines and small frontal area.

rougher running for a few hundred revs. above 3,000 r.p.m., after which the engine becomes as smooth as ever right up to and even beyond the red line on the rev. counter at 5,000 r.p.m. The exhaust note is (from the driver's seat) more noticeable in town than at higher speeds, when it is left behind. In fact, mechanical noise is very reasonably low, and it is the more disappointing that the metal hardtop and sliding side windows tend to magnify noise from the wind and the road surface to an unpleasant degree. The Triumph is, of course, available with a folding hood, which in addition to being quieter offers slightly more headroom. Two tall members of our staff were bounced into contact with the inside of the hardtop on separate occasions.

Economy

Noisy or not, this is a car where the driver tends to make full use of the performance available, if only because of the free-revving engine and the delightful gearbox. Fuel consumption suffers rather as a result yet, as steady-speed figures show, it is possible to be remarkably economical by restraint in gear-changing, without undue dawdling. A clear 39 m.p.g. can be obtained at 60 m.p.h. if the overdrive is fitted, and even without this unit fuel consumption figures are exceptional. The TR3 makes no demand for fuel better than British premium grade, and pinks only slightly on Continental "super" petrol. It should, incidentally, be noted that all performance and fuel consumption measurements were made with a pressure of 40 lb. in the tyres, at which the rolling resistance was naturally a little decreased.

High-speed cruising is favoured by long-range headlamps with a satisfactory spread of light. Visibility as a whole is good, the sliding side windows being less prone to scratching than earlier flexible types, while the wrap-around rear window leaves virtually no blind spot. Only the driving mirror could be better placed, at the top of the windscreen where it would not obscure the view of the near-side wing. Arguments can be advanced for and against the cut-away door shape, which undoubtedly admits draughts at the rear of the hinged fabric flap, but has the advantage, with the sidescreens removed, of greatly increased elbow-room for competition driving.

The excellent instruments, spoilt by a wildly optimistic speedometer, have been commented on already. Less confusion than might be supposed arises from the row of knobs for the minor controls, which include for the first time self-parking windscreen wipers. On the left of the facia is a locker of most useful size, openable only by key but extremely valuable in an open car. Unprotected carrying capacity includes a big space behind the seats (which can be occupied by a child's seat as an extra) and pockets in each door. Additionally, the lockable luggage boot will hold a large suitcase or a couple of canvas grips. Spare wheel, tools and jack are carried beneath it, the latter of improved pattern with a ratchet handle much easier to turn when the jack is fitted, through one of two holes in the floor of the car.

Routine maintenance is not yet a prerogative of the service station for many sports cars outside the luxury class. Low build and 13 greasing points do not encourage the owner to lubricate the chassis of the Triumph for himself, but there is good access to most salient points under the bonnet. The battery for one, mounted on the scuttle with slots in the bonnet for hot air to escape above it, is not likely to go short of water because of difficulty in reaching it. Like most aspects of the car, it leaves little to be added to the brief summary already attempted. A design with shortcomings but no vices, the TR3 offers a great deal at a modest price, and leaves a lasting impression of having been created by people who are practising motorists themselves.

Specification

Engine
Cylinders ... 4
Bore ... 83 mm.
Stroke ... 92 mm.
Cubic capacity ... 1,991 c.c.
Piston area ... 33.5 sq. in.
Valves ... Pushrod o.h.v.
Compression ratio ... 8.5/1
Carburetters ... Two inclined S.U. H.6
Fuel pump ... AC mechanical
Ignition timing control ... Centrifugal and vacuum
Oil filter ... Purolator full-flow
Max. power (net) ... 100 b.h.p.
at ... 5,000 r.p.m.
Piston speed at max. b.h.p. 3,010 ft./min.

Transmission
Clutch ... Borg and Beck, 9 in. s.d.p.
Top gear (s/m) ... 3.7 (overdrive, 3.03)
3rd gear (s/m) ... 4.9 (overdrive, 4.02)
2nd gear (s/m) ... 7.4 (overdrive, 6.07)
1st gear ... 12.5
Reverse ... 16.1
Overdrive ... Laycock de Normanville
Propeller shaft ... Hardy Spicer, open
Final drive ... Hypoid bevel
Top gear m.p.h. at 1,000 r.p.m. 20.2 (overdrive, 24.6)
Top gear m.p.h. at 1,000 ft./min. piston speed ... 33.4 (overdrive 40.8)

Chassis
Brakes ... Girling hydraulic (front, disc; rear, drum)
Brake drum internal diameter ... 10 in.
Friction lining area (rear) ... 86.8 sq. in.
Suspension:
 Front ... Coil and wishbone i.f.s.
 Rear ... Semi-elliptic
Shock absorbers
 Front ... Telescopic hydraulic
 Rear ... Piston-type hydraulic
Steering gear ... Cam and lever
Tyres ... 5.50-15 tubed

Coachwork and Equipment

Starting handle ... Yes
Battery mounting ... Under bonnet, on scuttle
Jack ... Ratchet screw
Jacking points ... Two, through floor each side
Standard tool kit: Adjustable spanner, two box spanners, tommy bar, tyre levers, three O/E spanners, grease gun, screwdriver, feeler gauges, tyre valve extractor, wheelbrace, starting handle.
Exterior lights: two head, two side/indicator, two tail/indicator, one tail/stop/number plate.
Number of electrical fuses ... Two
Direction indicators ... Flashing, self-cancelling
Windscreen wipers ... Electric, self-parking
Windscreen washers ... Optional, Trafalgar
Sun visors ... None
Instruments: Speedometer with decimal trip distance recorder, revolution counter, oil pressure gauge, water thermometer, ammeter, fuel gauge.
Warning lights: Dynamo charge, indicators, headlamp main beam.
Locks:
 With ignition key ... Ignition
 With other keys ... Glove box, boot
Glove lockers ... One, lockable
Map pockets ... Two
Parcel shelves ... None (large space behind seats)
Ashtrays ... None
Cigar lighters ... None
Interior lights ... None
Interior heater ... Optional, Smiths recirculating
Car radio ... Optional, Radiomobile
Extras available: Aero screens, overdrive, front undershield, wire wheels, rear wheel spats, leather upholstery, tonneau cover, heater, windscreen washer, telescopic steering column, Dunlop Road Speed or Michelin "X" tyres, aluminium sump, radio, two-speed windscreen wipers, competition front springs, competition rear dampers, fitted suitcase, occasional rear seat, fabric hood, "Grand Touring" conversion (fixed side screens and external door handles), 4.1/1 rear axle (when overdrive is fitted).
Upholstery material ... P.V.C.
Floor covering ... Carpet
Exterior colours standardized ... Seven
Alternative body styles ... Fabric hood

Maintenance

Sump ... 11 pints, S.A.E. 30 (summer), S.A.E. 20 (winter)
Gearbox ... 1½ pints, S.A.E. 30
Rear axle ... 1½ pints, S.A.E. 90
Steering gear lubricant ... S.A.E. 90
Cooling system capacity ... 14 pints (two drain taps)
Chassis lubrication: By grease gun every 1,000 miles to 13 points.
Ignition timing ... 4° B.T.D.C.
Contact-breaker gap ... 0.015 in.
Sparking plug gap ... 0.025 in.
Valve timing: I.O., 15° B.T.D.C.; I.C., 55° A.B.D.C.; E.O., 55° B.B.D.C.; E.C., 15° A.T.D.C.
Tappet clearances (cold):
 Inlet ... 0.010 in.
 Exhaust ... 0.010 in.
Front wheel toe-in ... 1/8 in. (Dunlop tyres)
Camber angle ... 2° positive
Castor angle ... Zero
Steering swivel pin inclination ... 7°
Tyre pressures:
 Front ... 22 lb.
 Rear ... 24 lb. (see text)
Brake fluid ... Girling
Battery type and capacity ... 12 v., 51 amp. hr.

ROAD TEST

An MT Research Report by Wayne Thoms

TRIUMPH'S VENTURES into the sports car field, dating back over 25 years, were notably unspectacular until the advent of the TR-2 with one interesting though short-lived exception—the Dolomite Specials of the '30s. Designed by famed Donald Healey, they featured a straight-eight engine with double overhead camshafts, a blower which boosted output to 140 bhp, and a guaranteed 100-plus mph.

Unfortunately, only a few were ever built and transportation vehicles pretty much made up Triumph's production, even after they were taken over by the Standard Motor Company in 1945. It was not until the London Show of 1952 that the first TR-2 was unveiled—a squat, businesslike roadster designed to be turned out in quantity at low cost. Enthusiast's appetites were whetted and two years later deliveries commenced in the United States.

Technical development of the TR-2 through the current TR-3 model has followed a course of careful, conservative evolution rather than revolution. Appearance, except for a flush-mounted grille, has changed very little. About the worst thing that can be charged against the Triumph is the fact that some consider its body lines uninspired. Aside from this—strictly a matter of personal opinion—a few hundred miles behind the wheel make you wonder why anyone could possibly want an engine larger than two liters displacement in a sports car.

In fact, Triumph's two liters (121.5 cubic inches) comes very close to being the ideal engine size in the TR-3. The overhead valve four, developing its 100 bhp at 5000 rpm, is not an especially modern design in comparison to some of the ultra-short stroke engines on the market. The engine is a warmed-up refinement of the Standard Vanguard mill with bore decreased through use of thicker wet cylinder liners to bring engine size into a more favorable racing class. Modern or not, one unalterable fact remains—it is one of the sturdiest, most trouble-free and reliable engines going today. It has plenty of low end torque, good mid-range performance and the ability to take long periods of high speed running.

Gear changes are crisp and firm with no feeling of excess play in the linkage. The short throw lever, operating remotely, sits atop the transmission tunnel—an easy, natural reach from the steering wheel. The optional Laycock-de Normanville overdrive ($160), acting on the top three ratios, offers a fascinating seven forward speeds which can be engaged or locked out without removing the hands from the wheel. A simple toggle

ACCESSIBILITY is keynote for engine compartment with most components in easy reach for minor maintenance.

switch on the left side of the dash makes this possible. Traffic will probably find you running all but the very slowest portions in direct third gear, switching into third overdrive when conditions warrant, making for a good deal of clutchless city driving. Aside from its novelty value—and it's being fun to operate—the overdrive becomes a wonderful highway cruising gear which boosts already excellent gas mileage into the phenomenal class.

Chassis and suspension remain little changed from the first TR's. Frame is a very stiff channel steel unit with an "X" crossmember. Front suspension is independent with coil springs and tubular shocks and rear consists of two semi-elliptic, longitudinal leaf springs. The most important single mechanical improvement has been the addition of caliper disc brakes in front. Disc brakes have a number of virtues, not the least of which is resistance to fade due to superior cooling characteristics. A series of panic stops from 60 mph just short of locking wheels produced smoke in amazing quantities from around the brake pads beginning at the sixth stop, some slight odor, a very little fade and a slight swerve to the left—all evils which would have been magnified greatly with conventional drum brakes.

In practical usage the discs are fade-free, new pads can be installed in a matter of minutes, and they are said to be self-cleaning and unaffected by water. This is certainly a boon to wet weather drivers and a fine no-extra-cost bonus on a medium-priced sports car.

Fit of body panels is uniformly good, as are paint finishes and bright work. Hood, deck lid and spare tire compartment just below the trunk open with a special "T" handle which one takes care not to lose. An integrated latching device, doing away with the loose handle for these three parts would be appreciated. The rug lined trunk compartment, spacious enough for a couple of small suitcases, locks with a separate key which also fits the glove box. Maximum utilization of trunk space can be assured through use of a fitted suitcase offered as a TR-3 accessory. Should there be occasion to tear into the engine, space and accessibility under the bonnet are very good. Incidentally, a chronic seeping of hydraulic fluid from the brake and clutch master cylinder on previous models has been corrected by redesigning the fluid reservoir.

As for interior, the TR-3 can probably lay claim to having more, or at least as much, legroom as any other sports car. Sliding the driver and passenger seats full back makes the car comfortably adequate for a pair of six-and-a-half footers. The leather-covered bucket seats offer fair support but can be tiring on long trips. On the passenger side, the seatback pivots forward for access to an extremely roomy package space behind the seats. An optional rear seat is available but we cannot recommend it

PERFORMANCE

Max. speed in gears, 1st 36 mph, 2nd 55 mph (o.d. 62 mph), 3rd 84 mph (o.d. 87 mph), top 107 mph (estimated). Acceleration: from standing start to 45 mph 5.4 secs., to 60 9.2 secs., ¼-mile 16.9 secs. and 81.9 mph, 30-50 mph 3.1 secs., 45-60 3.2 secs., 50-80 8.0 secs. Fuel consumption average for 441 miles 23.1 mpg. (101 miles on highway with heavy traffic, 60-65 mph, using o.d. 30.7 mpg.)

SPECIFICATIONS

ENGINE: 4-cyl. ohv. Bore 3.27 in. Stroke 3.62 in. Stroke/bore ratio 1.1:1. Compression ratio 8.5:1. Displacement 121.5 cu. in. Advertised bhp 100 @ 5000 rpm. Bhp per cu. in. .82. Piston speed @ max. bhp 3018.3 ft. per min. Max. bmep 145.8 psi. Max. torque 117.5 lbs.-ft. @ 3000 rpm.
TRANSMISSION: Hydraulically operated single dry plate clutch, 9 in. dia. 4 forward speeds, top 3 synchronized. O.d. available for top 3 gears—electric, positive. Overall ratios: 12.5, 7.4 (o.d. 6.07), 4.9 (o.d. 4.02), 3.7 (o.d. 3.03). Rear axle ratio 3.7:1. (Optional 4.1:1, available with o.d. only.)
CHASSIS: Rigid structure, channel steel pressings with "X" cross member. 5.50 x 15 tires. Brakes: Front—Girling hydraulic caliper disc, rear—drum with leading and trailing shoes. Cam and lever steering gear, with 34-ft. turning circle, 2.3 turns lock-to-lock.
DIMENSIONS: Wheelbase 88.0 in., overall length 151.0 in., overall height 50.0 in., overall width 55.5 in., minimum clearance 6.0 in., front tread 45.0 in., rear tread 45.5 in., weight 2200 lbs. (53% front, 47% rear), weight/bhp ratio 22:1.
PRICES (F.O.B. port of entry): Roadster $2625, hardtop $2790. (Slightly higher in the West.)
ACCESSORIES: Soft top kit (in addition to hardtop) $100, heater $40, wire wheels $111, chrome wire wheels $200, overdrive $160, adjustable steering $20, rear passenger seat $63, tonneau cover $35, Dunlop High Speed tires $24, competition shocks and springs (fitted at factory) $8, alum. engine sump $20, fitted trunk suitcase $48.

ROOMY COCKPIT offers honest sports car styling with plenty of hip and legroom. Bucket seats could give more support. Luggage space is small but of useful size and shape while spare rests in separate compartment.

TRiumph TR-3

except for small children—there just isn't adult size legroom in the back. A slightly wider door cutout would make entry easier.

Instruments are just where they should be and the dash is recessed to prevent glare. Speedo and tach are large and directly in front of the driver while oil pressure, gas gauge, water temperature and ammeter are centered on the dash. White numerals on black with good night illumination make everything readable. The large, locking glove compartment and two roomy door pockets offer plenty of incidental storage although why there is no ashtray or cigarette lighter, we cannot fathom.

One of the best looking sports car soft tops we have ever seen comes with the basic TR. Wraparound plastic windows make for excellent visibility, and the side curtains, with their sliding plexiglass windows, are as functional and sturdy as any on the market. The top, extremely weather tight, unsnaps completely and stores safely in the trunk while the bows fold neatly behind the seats.

The TR-3 also comes in a hardtop version, approximately $165 higher than the roadster. While the steel hardtop is removable with eight holddown bolts, difficulty of installing the fittings makes it impractical to buy the soft top version with the idea of purchasing the hardtop later.

The TR-3 handles beautifully at all speeds and exhibits no serious vices in the corners. There is a good feel of the road and the ride is firm though not unpleasant. A combination of short wheelbase and washboard roads will make for a choppy ride. In normal-to-hard dips there is very little rebound and oscillation, with no wallow at high speeds. It is a fairly simple matter to drift the car and the steering is about neutral, neither under- nor oversteer. Hard corners will induce more body roll than feels safe, but addition of the competition shocks and springs will stop this with a slight sacrifice in riding qualities. The engine sounds and feels smooth well past the 5000 rpm red-line—which most drivers exceed in competition—but they are strictly on their own. Piston speeds become dangerously high in this range.

As you may have guessed, we liked the TR-3. At the price, it is easily one of the best all-purpose sports car buys on the market. Acceleration, top speed, smart, precise handling, luggage space and driver-passenger comfort make it hard to beat. Along with this, an extensive dealer network and well-stocked parts warehouses make it a practical buy and fun to own.

◀ REMOVABLE TOP, good weather protection when up, stows neatly in trunk, is quite simple for one person to erect.

MOTOR TREND/SEPTEMBER 1957

CLEAN UNDERSIDE and good ground clearance of the latest TR are seen in this low angle shot. The intake grille is a smart feature of the new car.

TRIUMPH TR3

Disc brakes and over 100 b.h.p. in the latest version of one of the world's most popular sports cars

WHEN I performed the first road test of the then new Triumph TR2, I predicted a great success for it. My prophecy proved to be abundantly true, and this 2-litre sports car became one of the world's most popular speed models. Now, I have revived memories of that very first Triumph by testing its broadly similar successor, the TR3.

The TR3 retains the well-tried box-section frame, which is supported on wishbones and helical springs in front, and passes beneath the hypoid axle, from which it is suspended on underslung semi-elliptic springs. The engine now has larger SU carburetters, and is rated at 101 b.h.p. at 5,000 r.p.m. on a compression ratio of 8.5 to 1. It has a cast-iron block with wet liners and a three-bearing crankshaft, while the conventional pushrod-operated valves seat in an iron head.

The four-speed gearbox with remote control was one of the best features of the TR2, and this is unchanged. The optional Laycock-de Normanville overdrive was fitted to the test car. The disc-type front brakes were an interesting refinement. The two-seater body retains the bluff but not unattractive appearance which we have come to know so well, and is notable for its extremely useful luggage accommodation. Sports cars are often used for long-distance touring, yet they very rarely have adequate room for baggage. The Triumph must be particularly commended in this respect.

The hood, as always, is excellent, and the sidescreens are of a new and more attractive pattern. Entry and exit are not unduly difficult for such a low car. The seats are comfortable and have plenty of adjustment, while all the controls and instruments are well placed.

Hot or cold, the engine starts at once, and the clutch takes up smoothly as one moves off. The gearbox is just as delightful as always, and the indirect ratios are commendably quiet. I had to criticise the original TR2 for a noisy exhaust, but this has been eliminated, which greatly increases the pleasure of handling the car.

In matters of suspension and road-holding, I am raising my sights all the time. Thus, a car which called forth extravagant praise a few years ago might be regarded as being merely adequate today. This is natural as techniques improve, and so I make no excuse for subjecting the TR3 to a very searching test.

I would regard the Triumph as being a very safe car in the hands of the average driver. This is because it does not at first give the impression of holding the road particularly well. With experience, one finds that the adhesion is in fact better than it at first appeared, and the machine is always very controllable. Yet, for some reason which is not easy to define, it does not encourage the man at the wheel to take undue risks.

The cornering power is not exceptionally high, but when the limit is reached and the car slides, it remains perfectly easy to handle. The rear end does break away, but quite gently and with no undue tendency to spin. The rear semi-elliptic springs are commendably free from "winding up" during acceleration. The ride is definitely hard, but an acceptable degree of comfort is given on smooth British roads.

Thus, the Triumph forms a perfectly practical method of everyday transport, and with hood and sidescreens erect it would do very well for evening dress occasions. The car is flexible and has good traffic manners, and the lively

COMPACT dimensions and a neat, clean body style are evident. The Triumph manages to look equally elegant and functional—a rare combination—whether the hood is raised or folded away. Useful luggage accommodation is provided and the car is a practical form of everyday transport.

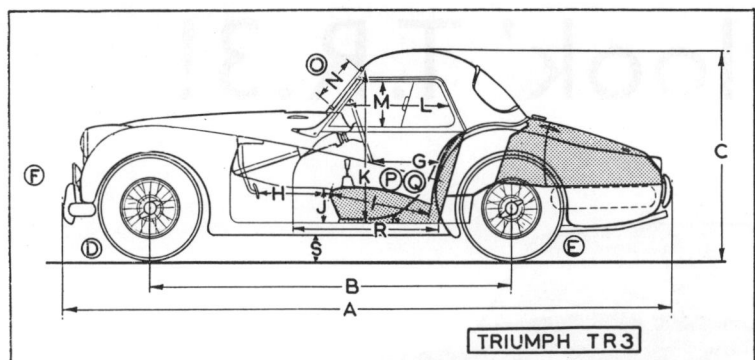

A Overall length, 12 ft. 6¼ ins.
B Wheelbase, 7 ft. 4½ ins.
C Overall height, 4 ft. 2½ ins.
D Front track, 3 ft. 10½ ins.
E Rear track, 3 ft. 10 ins.
F Overall width, 4 ft. 8½ ins.
G Squab to steering wheel, 1 ft. 2 ins. min.; 1 ft. 8 ins. max.
H Cushion to accelerator pedal, 1 ft. 3 ins. min.; 1 ft. 9 ins. max.
I Depth of seat cushion, 1 ft. 9 ins.
J Height of seat cushion, 8 ins.
K Height to top of screen frame from floor, 3 ft. 0 in.
L Width of side screen, 2 ft. 1½ ins.
M Depth of side screen, 10½ ins.
N Depth of windscreen, 11½ ins.
O Width of windscreen, 3 ft. 6½ ins.
P Width of seat cushion, 1 ft. 6 ins.
Q Overall width of seats, 3 ft. 9 ins.
R Door width, 2 ft. 1½ ins.
S Ground clearance, 6 ins.

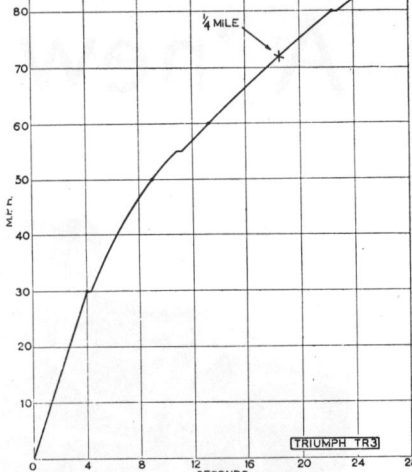

Acceleration Graph

acceleration is a potent safety feature in the right hands.

The powerful 2-litre engine naturally gives a very real performance to this small and relatively light car. A genuine, timed 100 m.p.h. can be exceeded on top gear, and on the overdrive a three-figure speed can also be attained, though the direct drive is a little faster. Third is a really splendid gear, and can be used right up to 80 m.p.h. for overtaking.

Curiously enough, this TR3 was fractionally slower than "my" original TR2. I have come to the conclusion that the earlier car was one of those lucky accidents—an exceptionally good one—whereas the TR3 was a more average sample. Fundamentally, the later model should have more power and therefore speed. In any case, the performance is all that any normal owner could desire.

I have spoken of the improved silencing. Another criticism that I made in my original road test concerned the brakes. The current car has the front discs which are now standard, and these really do overcome all the fading troubles. In the past, the rear brakes of Triumphs have not always stood up to their work, but some time ago the drums were increased in size, and all is now well. The man who drives on his brakes is a bad driver, but even he will be unable to overheat the "anchors" of the TR3. This is certainly the biggest single improvement that has been made available on the new car.

From an engineering point of view, this is a soundly constructed machine with nothing flimsy about it, and should stand up to plenty of hard driving. It is a straightforward vehicle for servicing, and provided that one has a pit or hoist, the general accessibility is quite satisfactory. Any owner who carries out the maintenance of an ordinary saloon will find the Triumph just as easy to keep in proper tune.

The TR3 is a fast and economical sports car, and by modern standards it is moderately priced. It has enough performance for competition work, but it is as a fast, long-distance touring car that it excels. Above all, it is a practical machine with good weather protection, adequate creature comforts and considerable luggage accommodation for a sports car.

FRONT WHEEL of the TR3 looks strangely "empty" because of the lack of a brake drum. Part of the disc-type brake is just visible.

Specification and Performance Data

Car Tested: Triumph TR3 sports 2-seater. Price £1,021 7s. including P.T. Extra, wire wheels £37 10s. including P.T.

Engine: Four cylinders 83 mm. x 92 mm. (1,991 c.c.). Pushrod operated overhead valves, 101 b.h.p. at 5,000 r.p.m., 8.5 to 1 compression ratio. Twin SU carburetters. Lucas coil and distributor.

Transmission: Borg and Beck 9 ins. single dry-plate clutch with hydraulic operation; 4-speed gearbox with short central remote control lever, plus electrically operated overdrive (optional extra). Ratios, 3.03 (o/d), 3.7, 4.9, 7.4 and 12.5 to 1. Short open Hardy Spicer propeller shaft. Hypoid rear axle.

Chassis: Box-section frame with cruciform, underslung at rear. Independent front suspension by wishbones and helical springs with telescopic dampers. Cam and lever steering, 3-piece track rod. Semi-elliptic rear springs with piston-type dampers. Centre-lock wire wheels (extra), fitted 5.50-15 ins. tyres. Hydraulic brakes, disc-type in front.

Equipment: 12-volt lighting and starting. Speedometer, rev.-counter, ammeter, water temperature, oil-pressure and fuel gauges. Flashing direction indicators.

Dimensions: Wheelbase 7 ft. 4½ ins. Track, front 3 ft. 10½ ins., rear 3 ft. 10 ins. Ground clearance 6 ins. Turning circle 32 ft. Weight 18¼ cwts.

Performance: Maximum speed 102.27 m.p.h. (direct top). Speeds in gears, overdrive 100 m.p.h., third 80 m.p.h., second 55 m.p.h., first 30 m.p.h. Standing quarter-mile 18.6 secs. Acceleration: 0-30 m.p.h., 4 secs.; 0-50 m.p.h., 8.8 secs.; 0-60 m.p.h., 13.2 secs.; 0-80 m.p.h., 22.4 secs.

Fuel Consumption: Driven hard, 25 m.p.g.

*

COCKPIT layout is neat and handy. Principal instruments are large and easily visible. The overdrive switch is visible on the right of the dashboard, handbrake lever is of the fly-off type.

*

A 'new look' T.R.3!

Price **£699 0. 0.** Plus P.T. £350 17. 0.

New T.R.3 features include: Redesigned radiator grille with built-in parking lights; new style front with recessed headlamps for smoother air-flow; new bucket-type seats with form-hugging back rest and attractive pleated upholstery; exterior handles on both doors for convenience and centre locking handle on boot lid for security.

Still way ahead—

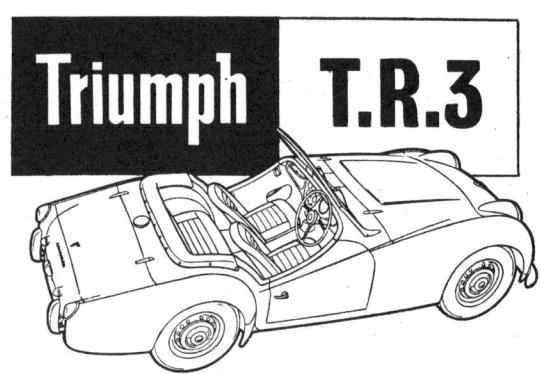

Your Standard or Triumph car is backed by a 12 months guarantee and the world wide Stanpart spares service.

Triumph Motor Co. (1945) Ltd., Coventry, England   *A subsidiary of The Standard Motor Co. Ltd.*   London Showrooms: 15-17 Berkeley Sq., W.1   Tel: GRO 8181

An Improved Triumph TR3

A New Front and Redesigned Seats

A RE-STYLED front, improved interior trim and external door handles (with locks) are the principal innovations which have been made in a new version of the Triumph TR3 Sports 2-seater which is now in production in both open and detachable hard-top forms.

As will be seen from the photographs on this page, the front panel between the wings and forward of the bonnet opening

The completely reshaped front and distinctive grille of the latest model are seen above while, on the left, the tail view shows the separation of the winkers and brake lights and the new external door handles. Production models will have a boot handle as well.

has been completely reshaped to accommodate a wide shallow grille which now embraces the combined side lamps and flashing direction indicators. The grille itself is slightly more forward than before and the faired headlamps protrude less, their lower portions in fact, now being slightly recessed.

In addition, the word "Triumph" now appears in spaced-out chromium-plated letters above the grille in conformity with the modern trend in name plates. The whole effect is both more modern and more attractive.

The separate bucket seats have more curved backs to provide extra lateral support and the trim generally has been improved. The central instrument panel is now black.

Stronger Over-riders

A detail of interest is that the upper portions of the bumper over-riders are now given additional support by stays which pass through the grille and are bolted to the inner wing pressings to provide additional strength for the sort of treatment bumpers are liable to receive in some overseas markets.

At the rear, the tail follows the former lines, but the word "Triumph" now appears in chromium-plated letters above the number plate and the lockable boot lid has a handle. The winkers, moreover, have been separated from the brake lights to avoid any possibility of confusion and the number-plate lamp is now plated. A further small difference in finish is the use of silver laquered wheels in place of colour.

So far as the interior is concerned, individual separately adjustable seats are retained, but the squabs are more rounded, bucket fashion, to give additional lateral support and the upholstery is deeper than before. Behind the seats, where there is useful accommodation for light luggage, the trim has also been improved to enhance the general appearance of the cockpit. The layout of the facia board is unaltered, but the central panel has a black lustre finish on the latest models.

Another change which will be very welcome is the fitting of external door handles so that it is no longer necessary to open one of the sliding panels in the rigid-framed Perspex side-screens in order to open the car when the hood and screens are in use. In addition, both doors lock with a key and the finger-grips on the outsides of the sliding Perspex panels have been eliminated to make it difficult to open the panels from outside the car when they are fully closed. Further, the interior lock pull-cords have been moved to an equally convenient, but less-obvious, position in the door pocket recesses. The result is to make the car much more secure against petty pilfering than most open models.

In other respects these new TR3 models conform to the 1958 specification which is already very familiar to readers.

MARCH '58

SPORTS CARS ILLUSTRATED

SCI ROAD TEST:

TRIUMPH TR3

SELDOM, IF EVER, remembering back over the last few years, have we ever met a man who bought a TR2 or a TR3 and regretted the purchase. We well remember the first time back in '54 when we climbed into a test TR2, one of the very first in this country. After a day with the car we were left wondering how they could bolt together that much car for so little money — it was one of those few cars that one is actually reluctant to clamber out of. Now, four years, loads of test miles and two models later we still get the same feeling — even more so. For sheer fun driving, the TR3 for '58 is hard to beat regardless of price.

The paramount changes in this model are in styling. The new latticed grille opening is recessed into a suggested snout-effect, a la Ferrari. The headlight bulges are smaller, and are also incorporated into the theme of the car, and of course are sealed beams. Across the hood the name is spelled out in large (but not too large) letters. A fuller and sturdier bumper spans the front, protecting the headlights and fenders as well as the grille. The appearance is a lot smoother because of these changes.

The TR3 supplied us by Standard-Triumph Motor Company was not a super-tuned cream puff. Service manager Peter Snow felt that the best way to evaluate a Triumph is to test the one that the next customer would have bought, so he just drove one out and gave it to us. This one is now a demonstrator.

But drive it we did! When we picked up the car, the odometer read *thirty-five* miles. Before making performance runs, or road tests of any kind, we just drove, putting over two thousand miles on the car in two weeks. This mileage ranged from close New York City traffic to ranging up to Belleayre Mountain on the Thruway for a week-end of skiing. The car behaved no matter what we did to it, averaged 26 mpg for the first thousand, and is now delivering in the order of 28 mpg. Unquestionably it will keep getting better. So far we've added no oil.

TR3 corners flat at high speed, such as on the "S" turns at Lime Rock. It seems to have an oversteering quality that makes it want power in turns to hold comfortable clip angles. Interior is roomy, well instrumented, finished in leather.

Engine is identical to last year's unit — one hundred very active horses that are easy to get at. Top speed: 104 mph.

A sports car for winter sports, too: in the interests of reader information, Associate Editor and friend borrowed a ski rack from Alpine Ski Shop and test-drove to Belleayre Mt.

The engine, as well as the gearing and other mechanical components, is identical to last year's Triumph engine, exactly — right down to the last bolt. There is plenty of power, even around 2000 revs, but the engine likes to *go* over the 3000 mark. This is the zone, 3000 to 4500, where you really move out when you punch the throttle.

The gear box on our new car was tight, but by the end of 2000 miles it slipped easily from gear to gear. The hydraulically-operated clutch is easy to work, gradually engaging, and positive when it pops in. But the short gear shift lever is perhaps the nicest feature in this department: we changed gears by reaching out and taking hold of the rubber dust cover on the stick, and changed gear slots by moving just the thumb. It's as easy as that.

Acceleration can be neck snapping if you want it to be: on the other hand the smooth-engaging clutch and good torque characteristics at the low end permit gentle take-offs, too. Two thousand revs is forty mph in fourth, but this same 2000 is plenty to get you off the mark, if you want smoothness and aren't in a hurry.

The brakes leave absolutely nothing to be desired. We made more than ten consecutive stops from sixty miles per hour — hard stops with just enough pressure to keep from locking the wheels. The adjustment of the rear (drum) brakes was faulty, and despite the fact that on every stop the right-rear wheel locked, our gauge reading was in the order of 2/3 "g," or approximately 70% efficiency on *every* stop. Apparently disc brakes of this type and size are able to do most all of the stopping. It is interesting to ponder, however, how we would have stopped if the brakes had been adjusted perfectly! The brakes felt as good when we finished as they did when we started.

The very first day that we had the car, it snowed. Taking the car out that evening, we were negotiating a twisting, unlighted, and deep-slush covered road at a fairly good velocity. The Triumph feels good even under these driving

Continued on page 90

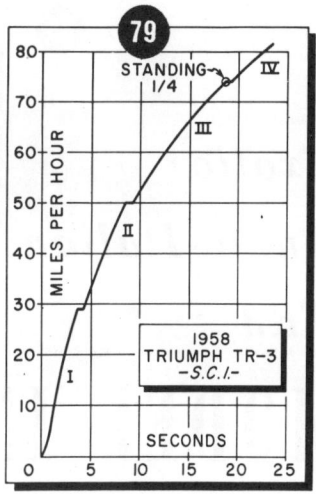

PERFORMANCE

TOP SPEED:
Two-way average .... 103 mph
Fastest one-way run .... 104 mph

ACCELERATION:
From zero to   seconds
30 mph 4.4
40 mph 6.3
50 mph 8.4
60 mph 12.6
70 mph 16.6
80 mph 22.6
Standing ¼ mile 18.6
Speed at end of quarter .... 74 mph

FUEL CONSUMPTION:
Hard Driving .... 17 mpg
Average Driving .... 30 mpg

BRAKING EFFICIENCY:
More than ten consecutive emergency stops from 60 mph were made at 2/3 of a g without any loss of pedal. However, on each stop, the right rear brake locked.

SPECIFICATIONS

POWER UNIT:
Type .... In-line 4
Valve Arrangement .... push rod ohv
Bore & Stroke .... 3.27 x 3.62 in (83 x 92 mm)
Stroke/Bore Ratio 1.11/1
Displacement .... 121.5 cu in (1991 cc)
Compression Ratio 8.5/1
Carburetion by .... Two Su H.6 sidedraft
Max. Power .... 100 bhp @ 5000 rpm
Max. Torque @ rpm .... 118 lb-ft @ 3000 rpm
Idle Speed .... 800 rpm

DRIVE TRAIN:
Transmission ratios I 3.38
                II 2.00
                III 1.32
                IV 1.00
Final drive ratio (test car) 4.11
Final drive ratio with OD ..... 4.55   (3.7 in OD)
Axle torque taken by .... Leaf springs

CHASSIS:
Wheelbase .... 88 in
Front Tread .... 45 in
Rear Tread .... 45½ in
Suspension, front .... Coil and wishbone
Suspension, rear .... Solid axle, leaf spring
Shock absorbers .... Telescopic front, piston rear
Steering type .... Cam and lever
Turning diameter .... 38 ft
Brake type .... Girling 11 in disc front
Brake lining area .... Girling or Lockheed drum rear
Rubbed area .... 248 sq in front, 87 rear
Tire size .... 155 x 15 Michelin X
                (equiv. to 5.50 x 15)

GENERAL:
Length .... 151 in
Width .... 55½ in
Height .... 50 in
Weight, test car .... 2135 lbs
Weight distribution, F/R 53/47
Weight distribution, F/R,
  with driver 51/49
Fuel capacity .... 15 U. S. gallons

RATING FACTORS:
Bhp per cu. in. 0.82
Bhp per sq. in. piston area . 2.99
Torque (lb-ft) per cu. in. .. 0.98
Pounds per bhp — test car .. 21.4
Piston speed @ 60 mph .... 2065 fpm
Piston sped @ max power ..... 3010 fpm
Brake's rubbed area per ton . 315 sq in

Exotic Anglo-Italian TR3

MODEL IN MANY RESPECTS

Above: Normally folded right away, the soft hood offers good occasional weather protection. Note the styled-in, built-in bumpers.

Left: Mildly finned and dartlike—styling borrowed from aviation—the Michelotti/Vignale coachwork includes no fantastic or impractical features. Minor point of criticism is the sharp-edged, difficult-to-clean brow over each head lamp

WHEN I visit the various continental motor shows year by year and see the variety of attractive special bodies made for foreign cars of all sizes, I want to kick the British governments which have just about taxed our equivalent native craft and design skills out of existence.

Special coachwork designs and fashions are a highlight at Turin, Geneva and Paris, but at Earls Court they have been pushed more and more into the background and confined to a few very expensive examples on even fewer exclusive chassis. This is a great handicap to our export efforts, and a loss to our home industry and market.

Having said this much, I can go on to talk of a rare bird, the British sports car with an Italian body. One or two Jaguars, Astons and others have, of course, been so equipped in the past year or two, but only a very few.

Seldom can a car have attracted so much attention as does the works Michelotti/Vignale/Triumph TR3, now nearly a year old and normally used by the manufacturer's managing director, Alick Dick. Anyone who has the pleasure of using it for a few days needs to get used to continual quizzing. All kinds of people stop and ask: Is it the new TR? Is the suspension all independent? What is the engine like? Is the hardtop detachable? are they Borrani wheels? Who made it? Are the gear ratios standard?—and so on.

Some of the answers are that except for details, it has a standard TR3 chassis, and is specially bodied for study in the TR development programme. It is a one-off design exercise, not a prototype as such.

In critical mood, I can seldom sit in a car or drive it without thinking "If it were mine, this and that would have to be changed for a start". After two weeks' regular use, the Michelotti TR seems to have more right and less wrong about it than any special sports car I can remember.

First, and briefly, the engine and chassis. The performance is all offered a little more sweetly than on the standard TRs I have sampled. Acceleration is very brisk and maximum speed is at least 100 m.p.h. The Laycock overdrive gives useful intermediate ratios between second and third, and third and top, as well as allowing a restful high cruising speed. The exhaust note is subdued. Plenty of sealing and sound proofing keep noises out of the car, and the engine is flexible and smooth for a sports four-cylinder unit.

Disc brakes are fitted at the front only and pedal pressures are on the high side, but the driver never has cause to worry about stopping. In this connection the Michelin X tyres provide excellent grip in all circumstances. They heavy-up the steering at manoeuvring speeds but not sufficiently to matter, and the floating sensation they introduce at higher speeds is no longer noticed after an hour or two of driving.

Road holding is particularly good and the rear end behaves well on fast corners—no sliding or hopping. But ground clearance is at least 2in too low.

Alfredo Michelotti and Giovanni Vignale have respectively designed and built a very practical body as well as an attractive one. It is some 200 lb heavier than the standard model, but this weight includes a number of extras and the detachable hardtop, in addition to a hood. It seldom pays to say a car is beautiful or ugly, because opinions differ so much, but there is no doubt that most people find the lines very attractive in this case. The workmanship and finish are admirable, and the cost, one off, ... but let's not go into that now.

An ample doorway avoids getting-in-and-out-troubles. The car is in black and white, both inside and out, relieved only by the leopardskin panels. The door-pulls lie flush and the driver's window winder folds flat

Unrecognizable as a TR3 but without a doubt most attractive, the car is here seen in the hardtop condition

The points which appealed to me in particular were the very good view from a compact, enclosed cockpit; the thin screen pillars, angled to give their minimum dimension across the driver's line of vision; then the boot, for a sports car, is very large and uninterrupted, lined stowage space being provided right across into the rear wings. So much space cannot have been easy to find because an 11½-gallon tank, the spare wheel and folding hood have to be stowed away in the tail as well.

Although the car is conveniently small, the cockpit is comfortably spacious for two, with plenty of leg room and enough clearance for head and elbows for a large driver.

To upholster the clutch and gear box hump and develop them into a console to carry ashtray and switches is a novel idea in this class of car. The leather-covered angle beside the driver's knee could be more rounded or padded with advantage. The diminutive gear lever is carried neatly and its knob is never more than a hand-span from the steering wheel rim. Very slick changes can be made, and the pedals are lined up conveniently and at a comfortable angle.

Being an exhibition showpiece, this TR is intentionally on the fancy side. I do not think a production version would have real leopardskin inlaid into the cream-piped black leather of seats and doors. The fur feels crisp, but is hot and rather slippery. The seats themselves are large, impressive but not very good. They have no wrap-round or shaping to hold you when cornering, and they are thinly padded beneath.

Grouping of the main instruments on a separate cowled panel in front of the driver, in the modern fashion, is to my mind right. For those who want to know the atmospheric conditions, an altimeter, hygrometer and air temperature gauge are neatly placed in the centre. I take a childish delight in a well-designed, self-extending and retracting radio aerial which whines up or down when the radio is switched on or off. The tone of the radio on this TR is good and the set is very selective. The volume control is rather too sensitive.

The wisdom of building the bumpers in with the body—a common practice these days—seems questionable. They are rigid enough to take a gentle tap but a good bump must surely be transmitted to the body panels, with the risk of expensive local deformation. There is no doubt about them being neat.

A sports car with a detachable hardtop is an excellent compromise. For weeks at a time the hardtop may be hung in the roof of the garage, while in the winter it stays on and gives the snug interior of a saloon. I never have liked side curtains—a winding window is always to be preferred. So there are two possible shortcomings to some people's way of thinking: the substantial, framed windscreen when the car is open, and heavier, thicker doors to accommodate the winding windows. The Michelotti car has no pockets in the door, but a good map container is located on the left of the passenger's floor.

Sun visors remain a problem on convertible cars. If they are to be big enough to do their job well they are a nuisance and have no place when the top is down. Rigid, transparent materials have their advantages but in a shunt might slice off the top of your head. There used to be roller-blind type visors. Perhaps the best answer will be a detachable visor with safely rounded attachment fittings.

This car was lent for sampling, not road testing, and to say much more would be to whet appetites that cannot be satisfied. But this is not to suggest that Triumphs will neglect the lessons to be learned from such a "special." In fact, there is no doubt that even now the Company is cooking up ideas and developing design features to be incorporated progressively in TR3 successors.
M. A. S.

Left: A very good boot for a sports car, wide, uninterrupted and neatly trimmed. The spare wheel is beneath the floor. Right: Cockpit—the driver's instruments are grouped directly in front of him, the tiny gear-lever is just under his hand, the screen is moderately and attractively wrapped-round, the radio is neatly built in, there is plenty of leg room and space around the pedals. The steering wheel might be an inch smaller in diameter

The new grille on the TR.3 is a notable improvement. The headlights have been lowered and the bumpers strengthened

TOURING TRIAL No. 8: TRIUMPH TR.3 HARDTOP

BRILLIANT performance, excellent handling qualities and a good measure of fuel economy are three of the requisites of sports cars today, and all of these features can be found in the Triumph TR.3 hardtop which we tested recently. The Triumph was at one time the "poor man's" sports car, but the ever upward spiral of prices has now placed it in the over £1,000 category; in fact the hardtop model sells at £1,102 7s. with Purchase Tax. It is worth mentioning, however, that even at this price it is still the cheapest of its type in the 2-litre class.

The test car had several extras fitted and these included a Laycock-de-Normanville overdrive which functions on second, third and top gears; knock-on wire wheels and leather upholstery. One useful item which was not fitted and is not even listed as an optional extra is an ashtray.

An all-steel body of clean and purposeful appearance is mounted on a cruciform braced chassis and the detachable steel hardtop greatly enhances the lines of the car. All wings and the complete front panel are of the bolt-on type and can easily be removed when required. The one piece, flat windscreen which, incidentally, is fitted with laminated glass, can be unbolted and aero-screens fitted instead.

The cockpit contains two most comfortable bucket seats which are adjustable for reach. The squab of the passenger's seat tips forward to give access to the small compartment behind the seats, which can be used for an overflow of luggage from the boot, or, if desired, to accommodate an additional seat. Entry is fairly easy, calling for no more contortions than are normally necessary to enter this type of car, in fact possibly less, because the door opening is wide and the seats are above the frame side members.

Both seats provide adequate comfort for long distance travel, are well shaped to fit the back, and prevent one being thrown about on a fast corner. The seats are now particularly attractively trimmed, and they are efficient. Should, by chance, the squab not fit snugly around the back and anchor the body, there is a grab handle on the facia for the passenger's use. Immediately below this handle is a capacious cubbyhole with a lockable lid; however, like most cubbyholes in cars produced by this company, the key is required to open and close the lid, which is occasionally a nuisance.

In the centre of the facia are the gauges for fuel level, oil pressure, coolant temperature and dynamo charge, the minor control knobs and the warning lamps. The arrangement is clean and sensible with the switches in handy, yet not easily confused positions. Completing the facia, and easily read through the T-spoked, 17-in. diameter steering wheel, are the two large dials of the rev. counter and the speedometer with its decimal trip and total mileage recorders. To the right of these instruments is the fingertip switch controlling the overdrive and the switch for the heater fan.

Between the left leg and the transmission tunnel is the fly-off handbrake lever, whilst on top of the tunnel itself is the stubby, rigid gear lever of the four-speed box. The lever is exceptionally well positioned as are the steering wheel, pedals and dipper switch. In all, it is a most workmanlike driving compartment and the only real criticism which can be advanced concerns the distance of the pedals from the seat; unless the driver is 6 ft. or over, he has the choice of sitting close to the steering wheel or moving the seat back to allow more elbow room and not reaching the pedals.

The hardtop has Perspex sliding panels in the doors with signalling flaps beneath, and a large wrap-around window in the rear. It affords first class protection against the fiercer elements of the British weather.

Basic safety is undoubtedly the keynote of the Triumph's design and this is met frequently as the examination of the car continues. Take the brakes, as the first example. Here we find a Girling hydraulic system operating very effectively under light pedal pressures. On the rear wheels are cast iron brake drums of 10 in. diameter housing shoes of the one leading and one trailing pattern, and giving a friction lining area of 86.8 sq. in. On the front wheels caliper action disc brakes are fitted, which really come into their own when hard and prolonged braking is required. The TR was brought to a halt from 30 m.p.h. in 31 ft.—a braking efficiency of 98 per cent.

Another safety feature is the method of securing the top opening, rear hinged bonnet, by two carriage locks, in addition to the usual safety catch. Below this bonnet lies a well thought out compartment with the 1,997 c.c. overhead valve engine as its centre piece. The eye first alights on the glittering rocker cover but is soon drawn to the two semi-downdraught S.U. H6 carburetters which meters fuel from the A.C. mechanical fuel pump to the four wet liner cylinders. The TR.3 engine produces 100 b.h.p. at 5,000 r.p.m. and it has a compression ratio of 8.3 : 1.

The engine is very lively; there is a hard mechanical noise when the road speed rises above 60 m.p.h. which, incidentally, can be reached from a standing start in 9.4 sec. It is possible to register 50 m.p.h. in 6.5 sec.; 40 m.p.h. in 4.8 sec. and 30 m.p.h. will appear on the clock just after 3.0 sec. In each of the seven ratios the car gives a good account of itself:

m.p.h.	2nd	O/D 2nd	3rd	O/D 3rd	Top	O/D Top
20-40	2.8	4.1	5.3	6.0	6.3	8.6
30-50	3.5	4.2	5.4	6.1	6.8	8.7
40-60	–	4.6	5.4	6.4	6.5	8.7
50-70	–	–	5.3	6.5	6.6	8.8

General accessibility of the components is very good, all being conveniently placed and within easy reach. Convenience also applies when it comes to getting at the spare wheel or tool kit as these are housed in a separate compartment beneath the luggage boot and can be withdrawn without disturbing the baggage. The boot itself has a flat floor and is most capacious for the type of car. Between the luggage compartment and the driving compartment lies the 12-gallon fuel tank which has a large, central, fast filling orifice with a snap action cover. Overall fuel consumption was 26.6 m.p.g., which included all the tests and much hard driving; it will be seen that the tank capacity will give the TR a most acceptable touring range. The consumption figures taken at constant speeds were:

At 30 m.p.h. .. In direct top, 41.1 m.p.g.
Overdrive top, 42.5 m.p.g.
At 60 m.p.h. .. In direct top, 28.8 m.p.g.
Overdrive top, 32.0 m.p.g.

For fast travel, the overdrive is a great asset. Whilst the four normal speeds have well chosen ratios and with synchromesh on 2nd, 3rd and top and a nice gear lever, changes can be easy and swift, the three intermediate ratios of the overdrive permit the selection of just the right ratio for the circumstances. The advantage of close coupled ratios above the 60 m.p.h. mark is well known to the sports car driver, and this car has no less than four operating in that field. The maximum speeds obtained in the gears were: 2nd, 50 m.p.h.; overdrive 2nd, 60 m.p.h.; 3rd, 80 m.p.h.; overdrive 3rd, 98 m.p.h.; Top, 110 m.p.h.; overdrive top,

Seen on a Monte Carlo reconnaissance, the new TR.3 is already familiar in the United States

(RIGHT) Engine accessibility is first class. The two carburetters can be tuned without anything having to be dismantled. Access to the plugs and dipstick is also easy and the under-bonnet layout commands respect

FEBRUARY, 1958

Ample leg room is a feature of the TR.3 seen here with the Bowmonk Dynometer attached to the windscreen. There is a full set of instruments and the arrangement of controls is ideal. The overdrive switch is most convenient

Spare wheel and tools live in a separate compartment, leaving surprisingly ample accommodation in the carpeted boot

115 m.p.h. In fact, during the tests, the needle of the speedometer went off the calibrations which finished at 120 m.p.h. but it was found to be reading fast by approximately 8 per cent at 115 m.p.h. It was indicating speeds 13 per cent fast at 30 m.p.h. and 14 per cent fast at 60 m.p.h. The distance recorder also registered the miles too quickly, by 2.5 per cent.

For a car of such capabilities, good visibility is essential and both forward and rear vision is of a high standard—all four corners of the car are visible from the driving position—but the Perspex side panels give rather a misty view of the sides.

In some timed high-speed motoring we found that 60 miles across country could be covered at an average speed of of just over 55 m.p.h. in daylight. Another journey of 56 miles in the dark, returned an average speed of 50 m.p.h. Much of the credit for the high speed of the last run must be given to the excellent headlamps which have a main beam power of 60 watts. The other lamps are of standard wattage but some form of internal illumination, such as a map light would be appreciated.

In general, roadholding and handling qualities of the TR.3 are excellent and the firm, conventional suspension system of coil springs and wishbones at the front and semi-elliptics at the rear provides an acceptable standard of comfort. As the test car was shod with Dunlop Racing tyres, delicate throttle control was necessary if nasty moments were to be avoided on slippery surfaces. As a point of interest, road conditions during the test varied from dry, to wet and icy, and the weather from freezing to rain and fog. One of the most memorable features of the car, was the pleasure it afforded on a 120-mile journey at night, when the driver and passenger averaged 44 m.p.h. over streaming wet roads, snug and warm inside the TR, enjoying the rapidity with which the miles passed, and the low burble of the beefy engine in overdrive in top gear.

It is a most desirable car which, in addition to its obvious sporting uses, provides ample touring comfort for two persons. It is small and compact, presenting little difficulty in parking and is extremely well behaved in city traffic where its surprising top gear flexibility and outstanding acceleration are both much appreciated. ★

An important improvement on the new TR.3 is the locking handle for the locker lid, which thus obviates the broken finger nails and bad temper which the previous locking arrangement provoked. Good visibility all round is a first-class feature of the hardtop TR.3

Used Cars on the Road—124

1955 TRIUMPH TR2 HARDTOP

The beige hardtop blends well with the green colour and the line of the Triumph. Head lamps are powerful but the beam is set low

Basic price new		£670 0 0
Total price new		£950 5 10
Price secondhand		£695 0 0

Acceleration from rest through gears:

to 30 m.p.h.	5.0 sec.	20 to 40 m.p.h. (top gear)	9.2 sec.
to 50 m.p.h.	10.5 sec.	30 to 50 m.p.h. (top gear)	9.9 sec.
to 60 m.p.h.	13.4 sec.		
to 70 m.p.h.	19.9 sec.		
to 80 m.p.h.	27.1 sec.	Standing quarter mile, 19.2 sec.	

Petrol Consumption	29-37 m.p.g.	Date first registered July 1955
Oil consumption	2,000 m.p.g.	Mileometer reading 23,517

Provided for test by St. Margaret's Motors, Ltd., 95, St. Margaret's Road, Twickenham, Middlesex. Telephone: POPesgrove 9075.

As a general rule the bodywork of a sports car receives careful treatment and preservation, and it is to the mechanical condition that a prospective buyer should direct his particular attention, to ensure that hard driving, with perhaps some racing or rally history, has not caused premature wear. On this comparatively young Triumph TR2, however, both bodily and mechanical condition were well up to the high average standards that are to be expected after only 2½ years' use.

Introduced in October 1954, this model is the detachable hardtop version, and its practically unused sidescreens and black hood are in the luggage locker. Beige plastic is used for the facia, interior trim and the seats, and this is mainly clean and sound, with the exception of a small tear in the side of the passenger seat cushion. Brown carpets—fitted in the luggage space behind the rear seats, and over the gear box hump—are unmarked. There is little sign of wear on the black rubber floor mats and the beige hardtop itself is practically as new on the inside, and very clean outside.

Green suits this model, and the TR2's paintwork in this colour is in outstandingly good condition. There are no scratches or rust on the chromium, and the external appearance is fine.

Starting was good throughout the test, and from cold very little choke was needed. The engine temperature rises quickly within a mile or so to its normal 185 deg F, but until this temperature has been reached the engine is hesitant unless a little choke is used. Although smooth at low revs, the power unit develops little torque below 2,000 r.p.m. Above that speed it fairly bursts into life, and will run up to 4,000 r.p.m. before any appreciable mechanical noise is evident; it then sounds busy—but not alarmingly noisy—right up to the danger line on the rev counter at 5,000 r.p.m. The range from 2,000 to 3,500 r.p.m. provides an extremely lively performance without stress; and it is used a great deal in normal road driving. The gear ratios suit this perfectly.

Indirect ratios are quiet, though there is some rattle of the gear lever in third. The synchromesh is effective and the stubby central gear change is a delight to use. The clutch is smooth, and absorbs effectively the load of standing starts under full power. It does not release completely when the pedal is depressed, with the result that it is difficult to engage bottom gear quietly to start from rest. To avoid "crunching" the gear it is necessary to pause for several seconds with the clutch pedal fully depressed, or alternatively to stop the input shaft by slipping the lever quickly into the synchronized second gear position before engaging bottom.

Steering is as precise as it must be for such a high performance car; the control does not transmit road shocks at high speed, and it is not too heavy in manœuvring. Suspension is firm but well damped, and the ride is comfortable. Directional stability and cornering are reassuringly good. A fault of the car is that scuttle-shake is evident at about 60 m.p.h., and becomes increasingly annoying at higher speeds. This occurred over most road surfaces, and was sufficiently bad to induce a tremor in the steering wheel, and for the instruments to become difficult to read.

The brakes proved to be thoroughly effective, and to have the reserves of power necessary to slow the car surely and rapidly from the 80-90 m.p.h. cruising speeds of which the TR2 is capable. The fly-off hand brake is efficient and convenient to use.

At low engine speeds the heater would run cold, suggesting that there may be an air-lock or blockage in the pipe; St. Margaret's Motors say that this will be rectified before sale. Other accessories on the car are a Pye radio (which has sufficient undistorted volume to be heard above the wind noise and the exhilarating but embarrassing exhaust roar); two wing mirrors of which the right one was invaluable; a narrow beam spot lamp; and a windscreen washer.

A splendid feature of the TR2 is its comprehensive and neat array of instruments on the facia: these and all of the car's electrical and mechanical equipment were in efficient working order —with the exception of the speedometer, which was so erratic as to be almost useless. This fault was not rectified by cleaning and lubricating the inner cable.

All tyres are Dunlops, of which the two front ones and the spare are new; on the rear wheels they are about half worn.

It is so rare in this series to come across a used car from which the toolkit has not been rifled that one wonders what happens to them all. This car, with only a jack, wheel brace and starting handle, was typical.

Unless the Triumph is positively stark and spartan with the hood up, it is difficult to see in what respect—except, perhaps, in appearance—one gains from having the hardtop. Draughts enter the car in all directions, particularly from around the sidescreens and the base of the windscreen, and there are many squeaks, rattles and booms. Admittedly, thanks to the panoramic rear window, the all-round visibility is excellent for such a low car; but there are plastic quarter lights in the hood, so that visibility should be even better when the hood is fitted instead of the hard top.

Acquaintance with this TR2 gave the impression that it had been very carefully used by its previous two owners, and that its purchaser should not only obtain in full the highly commendable performance and fuel economy for which the model is noted, but additionally a sound car which should have a long life of trouble-free service ahead of it.

The facia cubbyhole and the boot are lockable, but the car cannot be locked from outside. The perspex windows are only slightly scratched, and the rear sections slide easily

1958 SPORTS CARS ILLUSTRATED

SCI ROAD TEST:
DISC BRAKED

Photos by Albert Prokop   Karl Ludvigsen

Moving at about 80 through a fast left bend, the Triumph exhibits a slight lean to the right and very good control. Top and curtains didn't drum or rattle excessively, only fault being a tendency to lift less than an inch above windows when at or near top speed.

THE ruggedest single step in our road tests is the ten-stop braking trial. From a dead stop or nearly so, the machine is run up through the gears at 75-80 percent maximum effort. At a genuine sixty per, the clutch goes out and the brakes come on — hard, but not hard enough to break traction. The observer notes the maximum decelerometer reading, and as the car rolls to a rest, first gear is snicked in and away we go again. Ten times in a row — with a fast car taking about as long as it does to read about it.

This may seem unfair, since a car with less punch gets more time for the brakes to cool between applications. We think it's fairer than allowing a standard cooling time, since faster iron will in practice make greater demands on its brakes, which should be tested harder in proportion. This means more to you, who just might buy one of these.

Production Triumph TR3's now have a braking system that sets them wholly apart from all other sports cars in their class, in this important respect. Usually the braking test is a chore, but this time we were really curious. First run was smooth — heavy pedal pressure but maximum stopping without a waver. Second was the same, like most other good cars. At the third we laughed — deceleration went *up* a point! It stayed up on the fourth stop, and went up yet again on the fifth.

Still no swerve, still smooth, but the pedal was down a fraction. The sixth was lower, a driver factor, since the seventh was back up again. Each time we restarted now, in the frigid winter air, clouds of vapor billowed from the front wheel wells. They were hotter than any test brakes have ever been, but the ninth stop was as firm and fast as ever. Pedal feel was softening, and there were traces of swerve on the tenth halt, which was still among the best we've recorded. Discs were crackling hot, linings smelled to high heaven, but the darn TR3 stopped anyway.

Flashback — Le Mans, 1955: Ken Richardson wheeled a mixed lot of Triumphs up to the tech inspection before the 24-Hours. His own TR had Girling disc brakes at the front, and 11 by 2¼ inch Girling drums at the rear. The Dickson-Sanderson car had Dunlop discs all around, while Leslie Brooke herded a third car with standard Lockheed drums. Since they weren't after any out-and-out honors, this was extremely good experimental procedure, since Brooke's near-standard car acted as a control for evaluation of the two alternative braking systems.

In the race, the Dunlop discs had a negligible margin over the 15th-place Girling disc-drums, both being miles ahead of the old rig. On performance there was little to choose, so the Girling combo was picked on the basis of cost. It was also easier to hook up a handbrake using the rear drums, which were cut back to 10 inch diameter.

TR3

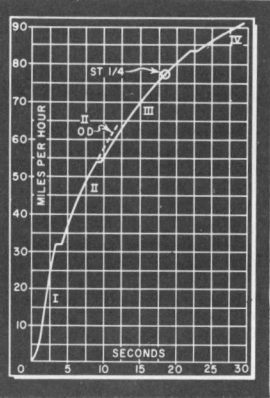

Modifications to intake ports don't show outside, big log-type manifold remaining the same. Battery is easy to reach, like junction boxes, water filler and heater valve.

If you buy a TR3, you'll become very familiar with this tool. It's here caught in the act of unlatching one of the two hood Dzus fasteners. Not very handy, but positive.

Arrow in this factory photo shows the single bleed fitting for twin spot cylinders. Removal of the two bolts and plates at the left allows linings to be pulled out by hand. Pads are at trailing side of disc.

As disc brake users are woefully discovering, good performance at Le Mans, where the brakes go on *hard* two or three times per lap, doesn't always mean impeccable braking on short courses. Discs do dissipate heat well, and have plenty of time to do so at the French course. They have little mass to store heat up, though, if two corners in quick succession keep them from cooling off. SCI's braking test is an extreme of this, and the refusal to fade that we encountered can be laid to the tough Ferodo DS1 linings, plus the fact that discs expand into and not away from the lining surfaces. Cooling had little to do with it.

Detractors of spot-type disc brakes have assured us that they'd get wet and dirty and cease to work at the slightest provocation. We wondered very much about this, and were overjoyed when we got snow, rain and slush for the test. Conclusion: The older two-leading-shoe drum setups were very tricky when wet; they'd either lock or not come on at all, or a combination of the two. With the discs, braking force remained perfectly proportional to pedal pressure.

Weather just didn't enter into it. If the discs did get really drenched, the first tap on the pedal would wipe them clean. Much credit for this must go to the leading and trailing edges of the lining segments, which are cut along radii of the disc circle. So don't let that worry you!

Eleven inches in diameter, the cast iron discs have a braking path 2¼ inches wide on each side. A husky casting

Lines are clean, square from rear, and good vision can be appreciated. Spare and tools are behind plate between twin rear "bumpers." Fuel filler is central.

Trunk access depends on that key too, with a central lock for added safety. Compartment is roomy, carpeted, here holds crank and tonneau cover (extra).

Big grab handle is handy for getting in and out, but deadly from safety angle. Lighter and ashtray are afterthoughts, other controls being well placed.

Taking a set for a tight right-hander, the TR3 cocks over but keeps all wheels square on the concrete. It will sliae controllably under these conditions, and is about to here. The Dodge van in the distance was a spectator — no legal motives!

embraces each one, and carries two chrome-plated pistons which actuate two segmental lining pads. The steel plates that form the backing for these pads have small ears which are held in place by triangular retainers. Removal of two bolts allows the segments to be pulled, and they can be inspected for wear through the access space. After each application the piston seals withdraw the pads to give .003 inch clearance. Car price has been held to that of the previous model by this very simple design, and Girling makes only the modest claim that these are "four times as effective" as drum brakes of equivalent diameter.

Brake and clutch pedals are suspended type (the clutch being hydraulically controlled), being hung from two cylinders high on the firewall. In older versions these had two separate fluid reservoirs, but service has now been simplified by fitting one can to feed both cylinders. This was also required by the relocation of the windshield wiper motor, which is now on the left-hand side. It's single-speed, self-parking, and its new design has made it almost inaudible, which was a pleasant surprise.

All this attention to brakes must mean that the TR3 can go as well as stop. It was fast to start with, and each year has seen a little more under the hood. Again Le Mans has helped out, the 1955 cars having specially shaped and inclined intake ports to suit the bigger carbs better. This is now production specification, and it seems to have smoothed out bottom end performance a lot. In contrast with earlier TR3's, the '57 engine will lug cleanly from low speeds and can be driven away quickly after a cold start (not that this is a good idea in any car).

Though the throttle linkage isn't rock solid, the engine itself is very responsive and has a wide range of power. There's a typical pushrod buzz, and some vibration periods, but exhaust is an unobtrusive hum. You won't antagonize the law with this Triumph. It has plenty of punch up to and including an easy 5500 revs, and never overheated or misbehaved.

The clutch can handle this power under all but the most extreme usage. We got very slight slip near the end of the performance trials, which will be worth it to many to get such smoothness in town use. Pedal throw is long and vague, but engagement is consistent and chatter-free.

As always, the TR3 gearbox allows full use to be made of the engine's good power range. Shift pattern is very compact with effortlessly short throws, and a stiff, stubby gear lever does the job. It's placed right where you want it, working through a conventional pattern with the right-hand reverse reached by lifting the lever. With good, though beatable, synchromesh, moderate gear noise and a good selection of ratios, the insides of the box are well planned too.

This isn't all in this department, since the Laycock overdrive tosses in three more ratios (it doesn't work on low — that would be pretty useless, and would put too much torque through the overdrive). Still very handy on the left side of the dash, the OD switch now operates more logically, in our opinion: Up for OD, down for direct. Engagement of the extra ratios is instant, with no throttle interlinking. There were no slip problems. Technically, third OD is the only superfluous gear in the lot, and even it has its purposes when used in sequence with third direct Second and second OD are a good pair for town use, giving

All in a bunch on the left hand side are the clutch and brake cylinders and wiper motor. The latter is excellent, being potent and silent. New single fluid reservoir is clamped at left.

you an available top of 65 mph. Overdrive top, of course is pure luxury, allowing you to cruise virtually flat out. As usual, since it's way overgeared, top speed in OD is slightly less than that in direct. With the optional 4.1 to 1 rear-end gearing this could change for the better.

Like the gear controls, the rest of the knobs and levers are easy to get at and positive in action. Out of the driver's way on the other side of the tunnel, the lever-type hand brake is still easy to reach and very effective. Brake and clutch pedals are firm and distinct, though their leftward placing can cause confusion at the very first. The left foot rests naturally on the big dimmer switch. A minor point is that the early floor carpeting has been replaced by rubber mats, which are less likely to bunch up and snag the driver's heels.

Electrical controls are grouped handily at the center of the dash, sharing space with the four secondary instruments. These, like the big tach and speedometer, are well marked in white on black faces, and well lit (at one intensity) for night work. There's no separate interior light, but some scattered light under the dash will help out. The speed on our car was unusually accurate, and has an adjustable trip mile-counter.

Thanks to the use of side curtains, the TR3 has both a big, lockable glove compartment and handy map pockets in the doors. There's also a lot of space just behind the seats, where the optional kiddy-type bench can be installed. Here also the trim has been simplified. A smooth, board finish is used instead of carpeting as in the TR2, giving a much neater appearance. Interior trim in general is rugged and livable.

(Continued on page 90)

TOP SPEED: | 4th | 4th OD
Two-way average .... 104.1 | 102.1
Fastest one-way run .... 105.9 | 104.7

ACCELERATION:
From zero to Seconds
30 mph 3.4
40 mph 5.7
50 mph 7.9
60 mph (1, 2, 3) 11.5
60 mph (1, 2, 2OD) 10.9
70 mph 14.9
80 mph 20.0
90 mph 28.4
Standing ¼ mile 18.0
Speed at end of quarter ... 76.5 mph

SPEED RANGES IN GEARS:
I .... 0 to 32 mph
II .... 4 to 54 mph
II OD .... 8 to 65 mph
III .... 10 to 83 mph
III OD .... 15 to 102 mph
IV .... 20 to top
IV OD .... 24 to top

SPEEDOMETER CORRECTION:
Indicated — Actual
30 — 32
40 — 41
50 — 50
60 — 59
70 — 69
80 — 79
90 — 89
100 — 100

FUEL CONSUMPTION:
Hard driving .... 16.3 mpg
Average driving (under 60 mph) . 25.5 mpg

BRAKING EFFICIENCY:
(10 successive emergency stops from 60 mph, just short of locking wheels)
1st stop 70
2nd stop 70
3rd stop 71
4th stop 71
5th stop 72
6th stop 69
7th stop 71
8th stop 71
9th stop 71
10th stop 69

SPECIFICATIONS

POWER UNIT:
Type .... Four cylinder, in-line
Valve arrangement .... Overhead, in-line
Bore & Stroke (Engi. & Met.) . 3.27 x 3.62 ins. (83 x 92 mm)
Stroke/Bore ratio 1.11/1
Displacement (Engi. & Met.) .. 121.5 cu. ins. (1991 cc)
Compression ratio 8.5/1
Carburetion by .... 2 SU H.6 sidedraft
Max. bhp @ rpm .... 100 @ 5000
Max. torque, lb.-ft., @ rpm .. 118 @ 3000
Idle speed .... 850 rpm

DRIVE TRAIN:
Transmission ratios
Rev 4.27
I 3.38
II 2.00
III 1.32
IV 1.00
Final drive ratio (test car) .. 3.7 (with 0.82 overdrive)
Other available final drive ratios. 4.1
Axle torque taken by .... Rear leaf springs

CHASSIS:
Wheelbase .... 88 ins.
Front tread .... 45 ins.
Rear tread .... 45.5 ins.
Suspension, front .... Coil and unequal-length wishbone
Suspension, rear .... Solid axle, leaf springs, underslung frame
Shock absorbers .... Telescopic front; piston-type rear
Steering type .... Cam and lever
Steering wheel turns L to L .. 2
Turning diameter .... 33 ft.
Brake type .... Girling hydraulic, 11 in. discs front
Brake lining area .... Rear: 87 sq. ins.
                               Front: (rubbed area) 248 sq. ins.
Wheel studs, circle diameter . Rudge splined hubs
Tire size .... 5.50 x 15

GENERAL:
Length .... 149 ins.
Width .... 55½ ins.
Height .... 50 ins.
Weight (test car) .... 2200 lbs.
Weight distribution, F/R 52.75/47.25
Weight distribution, F/R,
  with driver 50.6/49.4
Fuel capacity, U.S. gallons .. 14.4

RATING FACTORS:
Bhp per cu. in. 0.823
Bhp per sq. in. piston area .. 2.99
Torque per cu. in. 0.976
Pounds per bhp (test car) 22.0
Piston speed @ 60 mph .... 1800 fpm. (OD: 1470 fpm.)
Piston speed @ max. bhp ...... 3010 fpm.
Brake lining area per ton
  (test car) .... Using front rubbed area—304 sq. ins.

Triumph

Continued from page 89

Separate bucket seats are now placed at a comfortable angle, and have moderately curved backs which give fair lateral support to the occupants. As with most sports cars, getting in and out can be awkward with the top up, including the operation of reaching for the interior door latch cord. When you plop into the seat, though, you find more leg room than you can possibly use, plus plenty of head room. The one sore spot, just as bad now as ever, is the left elbow room with side curtain in place. With the car open the door cutaway gives you action space, but it lets in a lot of wind around the kidneys.

We liked the TR3 driving position very much, since the room inside and good seat angle are matched by a steering wheel of good size placed well away from the driver. This car had the non-adjustable column, which is actually better than the sliding version if you prefer this long-reach style of driving. Such isn't really required by the Triumph, which has steering fast enough to avoid extreme cross-hands maneuvers. In all respects the steering is par for a production sports car, having a slight amount of play, and good road feel with moderate kickback and self-return action.

Its slightly forward weight bias and inclined rear springs give the Triumph a moderate understeer characteristic. It thus tracks well on the straight — isn't tiring to hold — and has no vices at medium cornering speeds. When the show really gets rolling the tail end makes its presence felt, and the transition from tracking to sliding can be sudden and choppy. The margin of predictability is just wide enough to make the car enjoyable to the sports driving newcomer, though, who will be able to anticipate most of the car's motions.

Response to the wheel is quick, once the front tires take a bite. Triumph recommends tire pressures six pounds higher than standard for fast driving, and we adhered to these. Using standard Dunlops, the amount of tire noise on corners was negligible. Yes, it does lean, but not enough to affect control.

Cornering agility of the TR3 is partly a result of the short wheelbase, which also causes a bounding, pitching motion over some undulating road surfaces. The experimental boys in England have tried to mollify this by changing shocks and spring rates around, and the present combination is probably the best so far. With moderate tire pressures the TR3 gets over smaller road ripples very smoothly. It's a good compromise between ride and roadability.

Continued from page 79

TRIUMPH TR3

conditions. Within a split second a dog ambled into our path. We stepped on the brakes — hard — and the TR3 stopped *straight* in a distance that seemed no longer that if we had been on dry concrete. It's pretty hard to do better than that.

In order to determine just how well our TR3 handled, we then took it racing (against the clock) at Lime Rock. When we drove through the gate, we were somewhat perplexed to find that part of the track was covered with *snow;* nonetheless we had a crack at it.

The steering is very quick, and when cornering at low speeds there is a tendency (or a need) to straighten the car out a bit after it's been committed to a line. However as the speed picks up, this necessity seems to disappear, and tracking is quite easy. It's a stiff-feeling wheel, with no play and very little return, but it feels good regardless of vehicle velocity.

We ambled around the course a few times, and on one turn a combination of road ice and driver enthusiasm sent the tail out a little too far. Fact is, we spun. But the important thing is that we spun *flat* — it never even came close to going over, and we always had it under control even when it was out of shape. And despite the biting cold weather, the heater kept us comfortable.

The interior is finished in leather, with rubber mats on the floor. It's a lot easier to get into this TR3 than it was last year, because real, honest-to-goodness twist handles have been placed on the outside of each door (ever close the doors with both curtains snapped shut?). There is a large range of seat adjustment, enough to cater to anyone under seven feet tall. The seats are soft, bucket-type, placed so that there is plenty of elbow room.

And there's also room under the hood to get at and work on the engine. The plugs are in the open, as well as the SU's, carb linkage, battery, hydraulic fluid etc. When we completed the first of our high-speed runs over the SCI course, and turned around for the return, the engine developed a terrible miss. The Tech Editor raised the hood, located the trouble, and secured the hood within one minute. The trouble was in a carb dash pot that had loosened. He didn't even need a light.

There is one thing about the car that we complained about last year, and we will have to pan again. The exterior of the doors are curved surfaces, however the mating surfaces of the side curtains are flat. The result is that the surfaces meet only at the center, allowing cold air to channel onto the back of the neck. We made the car very comfortable by stuffing two wool mufflers into the gaps, however it seems a shame that Triumph couldn't either have curved the side curtains to the contour of the door, or installed a simple piece of insulating rubber.

On the other side of the ledger, the things that impressed us most was the excellent quality control at the Coventry works. The dash panel is fitted with finely made instruments, and you get the feeling that if you own a Triumph for a hundred years, nothing is going to fall off of it. This, unfortunately, can not be said of all our domestic automobiles.

Last year's TR3 sold for $2625; this year's sells for $2675. It would appear there is a price increase, but we don't think so: you get more automobile. The improvements and conveniences that were put on at the factory would cost more than fifty dollars, even the do-it-yourself way. And every one of them is worthwhile.

—Len Griffing

PRACTICAL CLASSICS BUYING FEATURE

Buying A TR2, TR3 or TR3A

Michael Brisby discusses a sportscar for real enthusiasts which offers unusual economy combined with performance, reliability and value for money.

Only the most ardent TR fans would suggest that the Triumph TR2 and its successors, the TR3 and TR3A, are perfect motor-cars but, on the other hand, who could deny that they were and are REAL sportscars? In their day they were fast and tough enough to perform very well on race circuits, they were very competitive rally cars — an outright win in the RAC Rally proves that — and yet they could also put up remarkable performances in economy events, winning the 1955 Mobilgas Economy Run outright with an incredible 71 m.p.g.

Nothing about the TRs was complicated or exotic. They were simple and robust, achieving their successes with a combination of stamina and good power to weight ratio. Best of all, they were good value for money.

Good Value

What has the TR2, 3, or 3A got to offer today? Obviously the cars have plenty of nostalgic appeal, they still have their very individual looks and all the original performance, reliability and economy is still available if the car is in good heart. In fact, if you set aside road-holding, handling and comfort a "proper" TR could give a TR7 a good run for its money and we were at one time sorely tempted to prove that point by carrying out a comparative test on the two cars.

However, there is more than looks and performance to recommend the TRs to *Practical Classics* readers. First things first — they are really not all that expensive to buy when an outstanding example can be bought for under £4,000, a runner from £800 to £2,000, and candidates for total restoration start at £300. They are cheap to run — there is something wrong if you cannot better 30 m.p.g. — and the mechanical components are very reliable and long-lasting. Restoration could not be easier because the cars are very simply constructed and parts availability is excellent.

History and development

The story of the Triumph TR sportscars has been well documented in several books by people who ought to know what they are talking about. Briefly, the story started in 1953 with the TR2 which was intended to trounce the MG TF and undercut the Austin Healey 100 at home and in vital export markets (principally the lucrative North American market.). The design incorporated a much improved version of the Standard four-

First of the line, the honest, no frills Triumph TR2 provided performance with remarkable economy and Reliability and those qualities ensure that the car is still a very attractive and practical proposition today. 8,628 TR2's were built between 1953 and 1955 and the recessed grille is an instant recognition feature.

The grille shape and position identifies this as a TR3 which benefitted from disc braking at the front. The factory offered an excellent hardtop and a pretty useless occasional rear seat. If you can keep the speed up a keen TR driver can stay fairly warm and dry with just a good tonneau cover for protection.

Buying A TR2, TR3

Continued cylinder wet liner engine which was proving its toughness in the Vanguard, the Triumph Renown and (as no one will allow you to forget) the petrol-paraffin version of the Ferguson tractor. Money for body tooling was in short supply but the very simple body shape had its own appeal and most owners were prepared to tolerate the car's playful tendency to "exit stage left backwards" because it went so well between corners!

The sales succes of the car added urgency to efforts to improve the car and its brakes were swiftly improved in performance and balance. The TR was a low car and it is said that the lower edge of the door was lifted to stop it hitting high kerbs and there is also the possibility that this improvement provided a convenient opportunity to adopt a different sill arrangement which made the body usefully stiffer.

The Triumph TR2 had its grille at the back of its air intake, just ahead of the radiator, and its successor, the TR3, had its grille at the intake entrance. The TR3 was introduced in October 1955 and apart from the grille and other minor cosmetic changes the power was increased by 5 b.h.p. to 95 b.h.p. with bigger inlet ports and 1¾ SU.s.

A further 5 b.h.p was extracted from the 1,992 c.c. engine when the "high-port" head was adopted during 1956 but the next obvious change involved a third alteration to the grille, making it much wider, and slight revisions to the headlamp pods. In addition the doors and bootlid (previously secured by carriage locks) were given external handles. In this form the car has become known as the TR3A although the badgework and brochures did not refer to it as such. Most customers were more impressed by the provision of disc brakes at the front. Incidentally the TR2 and 3 had very little chrome decoration but the 3A had the letters TRIUMPH pinned above the grille.

The TR3A coincided with or created a considerable overseas demand, and sales remained very high for over two years before it became clear that something new with more creature comforts was required; and that came in the shape of the TR4 in 1961.

Variants

Between 1953 and 1961 the current Triumph TRs provided the basis of important components for various cars. Morgan, who had been using the Standard Vanguard engine for some models took advantage of power improvements at the various stages of the same engine's development for the TRs. The Belgian Triumph importers converted the bodywork of some TR2s to create a fixed-head coupe with "wind-up" door glass which they called the TR2 Franco-champs after the Spa motoring racing circuit, while in Britain the Swallow Coachbuilding Company (sold off by Jaguar after the war) of Walsall produced the

Over-riders but no bumper at the rear of the car does not offer much protection. With only 6-inches of ground clearance the doors were altered after it was discovered that the lower edges were prone to ground on kerbs and it is only the early TR2s which did not have the sill showing below the door.

With cutaway doors, big wheel, short gear-lever and plenty of purposeful looking instruments the TR interior is everything one would expect of a traditional sportscar.

Swallow Doretti which lasted about eighteen months between 1954 and 1956.

The Swallow Doretti had a tubular steel chassis and used the TR2 power unit, transmission and running gear. The body followed the same styling fashions as the Austin Healey and MGA and it seems likely that it was this pleasant looking car's high cost rather than the fact that Sir John Black, Standard's Managing Director, had a rather major accident on one, that killed the project.

The Peerless and the Warwick were much the same thing and both had TR bits below their glass fibre surface, and Daimler's SP 250 (neé Dart) with its lusty 2½ litre V8 might have been a better bet if their engineers has copied the wider track, rack and pinion steering, TR4 chassis rather than the TR3A.

Then there is the Triumph Italia, probably the most appealing of the bunch. Michelotti who did a great deal of styling work for Triumph designed the steel body, but they were built in Italy by Vignale. The body

or TR3A

The very much wider grille, revised headlamp teatment and to a lesser extent the external door and boot handles identify this as a TR3A which was current between 1957 and 1961, and sold still better than its predecessors, especially in export markets.

Do not dismiss the four-cyinder Triumph engine as crude. It is powerful yet produces plenty of torque, it is tough, long-lasting and easy to work on and at a constant fifty in overdrive top does close to fifty miles per gallon. At 70mph the engine is turning at less than 3,000rpm.

looked like a TR4 frightened by a Maserati and was a fixed-head coupe, but it cost too much to succeed. The cars were built between 1959 and 1963 but very few came to Britain.

Lastly, just as the description Triumph TR3 is not official, so TR3B is even more unofficial and understandably confuses people. Some sources state that the TR3B was concocted by the factory to please American dealers who were not enthusiastic when they were allowed a preview of the TR4. I feel that deliveries to the North American market were interrupted as the TR3A went out of production and tooling for the TR4 was readied and that the TR3B perhaps filled the gap. That would explain to some extent why the first 500 had the 2,138 c.c. engine (latterly optional on the 3A) and the non-synchromesh on first gearbox was fitted, while the last 2,500-3,000 cars utilised the new, all-synchomesh TR4 box.It may be my ignorance but I have not heard of any body or trim details which make the "unofficial" TR3B instantly recognisable and it should be stressed that these cars were for export only.

You might buy a TR in this sort of state for between £300 and £500 but if the car has been dismantled make quite sure it is complete or that you can obtain replacement parts. A reproduction front panel for this car would be relatively expensive.

What to Look For
Continued

There are now all sorts of TR to choose from, regardless of whether you go to look at a TR2, TR3 or TR3A. They range from a few original cars, through properly rebuilt ones, to patched up but basically sound ones to the patchwork galloping wounded which show signs of hard use by generations of carefree individuals who appear to have taken a pride in proving that Land-Rovers are for sissies! The challenge is to decide which grade of dereliction you are looking at and due to the simple construction of the cars a thorough visual inspection will tell you a great deal.

A total rebuild of the engine, transmission and running gear is no more expensive than for any other car but if you have to do it at least you can be pretty sure that the car will not wear out for a very long time. The legend

Rust and accident damage are not uncommon at the rear of a TR. The vast majority of the panels are available but make sure you know what you are taking on before buying.

Buying A TR2, TR3

that the four-cylinder TR sportscars use a tractor engine is a misrepresentation of the facts (it is safer to say that the tractor used a Vanguard engine and farmers thought it a bit feeble!) and it is long-lasting, should be easy to start and also relatively smooth. A tired engine will probably have considerable wear in the valve gear and if the rings are worn (and I have seen them so worn you could see that they were oval) crankcase pressure will blow oil out of the timing chest and rear main oil seals. Unless you know better suspect the carbs of being worn and listen for a broken timing chain tensioner clattering at low revs. Incidentally, the TR engine will still go pretty well when it shows every other sign of being on its last legs!

Having the bonnet open and perhaps dazzled by a chrome plated rocker cover that looks as if it might have been a prize at a fair ground shooting gallery, cast an eye over the front inner wings. A combination of circumstances — unsightly and feeble front bumpers, and energetic driving over many years — can spell front end shunts. Rust along the edges where the wings bolt on and at the rear where the inner wing meets the bulkhead is not uncommon either.

The bottoms of the front wings may be rotten — provided that nobody has slipped on fibreglass or aluminium panels. If not feel the inside of the wing as far as possible to see if a repair section has been put in. Perhaps the sills of the TR2 and 3 do not do much but they should be sound — new ones are available so there is no excuse for patch repairs. The bottoms of the doors can rust but new skins don't seem to be available.

The whole of the rear section of the body should be examined for rust and accident damage. Over-riders but no bumpers do not offer much protection. Examine the rear panel which rots and take out the spare wheel to see how the floor of its stowage space has lasted.

TR wings bolt on and the front ones lend considerable strength to the front inner wings which can be attacked by rust. The floors and forward sections of the footwells and sill structure are likely to require replacement in the course of a rebuild.

TR's have stiff suspension and limited rear axle movement and you have to accept some chassis flexing as part of the fun. Rust attacks the body mounting out-riggers, the rear of the main chassis rails and the tubular cross-members.

Trim? Well I have always thought that Triumph trimmed the early TRs quite well and that the weather protection was also quite good. The works hard-top is an excellent device. While looking inside the car lift the carpets and look at the floors paying particular attention to the body mounting points which rust or split or both.

Hard usage and old age on a car with little ground clearance make it advisable to look at the car's underside. At the front it is accident damage and splits that you are looking for, further back it is grounding scars and from the seats back, rust at the rear of the chassis rails and in the cross tubes. You should also look at the tubular out-riggers which carry the body mounting brackets — the tube can rust away, and while it is interesting to know that the TR has a separate chassis it should **not** be all that separate.

The Peerless had much in common with TRs below the surface bit was not a commercial success, nor was its revival as the Warwick GT, seen here, any more fortunate.

or TR3A

It is not fair to accuse the TR of rusting badly, when their age and usage is taken into account a certain amount of decay is to be expected. Here there are signs that the rear wheel arch and lower body are both under repair and the bottom of the door has rusted in typical fashion. Original panel fits were good.

Driving a TR

If the early TR sportscars were meant to have been comfortable and easy to drive they would not have bothered with the 2 and 3 and would have given us the TR4 from the outset! You may find the steering wheel big and close but a few trips on rough-surfaced winding lanes will convince you that such an arrangement is essential.

The near vertical fly-off handbrake is pleasant to use. The instruments may not be works of art but they are clear and you can see them; the pedals are well placed. The seats on my brother's "long-door" TR2 gave a bouncy ride and with cutaway doors you are conscious of sitting very close to the road. I found that the lack of any noticeable suspension made throttle control a bit crude on rough roads and the car skitters about a bit.

The overdrive makes a TR, allowing 24.4 miles per hour per 1,000 r.p.m. in overdrive top and I vividly remember having accelerated hard to eighty and then engaging overdrive and after a few moments thinking how effortless the performance was, beginning to worry about things like corners and brake fade. Braking on the drum braked TRs is adequate if everything is in good order and that on the disc-braked cars should be good.

Do not attempt to corner quickly in the wet until you know your TR and be a degree or two more cautious if the car is on ancient radial tyres which can make the tail-end breakaway a bit sudden; a friend of mine proved that on a roundabout second day out!

The gearchange quality gives the impression that the gearbox design was intended for earth-moving equipment but on the other hand you won't miss a gear and the ratios are well chosen, once you appreciate that the engine develops its torque at 2,800-3,000 r.p.m. and does not need rowing along on the gear-lever.

Restoration

The first principle for TR restoration is to tackle the job in the correct sequence — repair or temporarily brace the body so that it is strong enough to be lifted off the chassis (a popular dodge is to carefully cut the body in two), repair the chassis, place the body back on the chassis and reconstruct it. A very large proportion of the parts needed for total restoration are available but there are not all that many professional restoration concerns who have the will or, apparently, the ability to do the work for you. However, the TR Register is bursting at the seams with those who know how to do it having carried out successful rebuilds of their own.

The pleasant looking Swallow Doretti used Standard Triumph bits common to the TR in its own chassis but was too expensive to succeed.

Clubs

The major club catering for the cars is the well-established and enthusiastic **TR Register**, membership enquiries should be made to Mrs Valerie Simpson, TR Register, 271 High Street, Berkhampstead, Herts. The recently formed **TR Drivers Club** has its membership inquiries dealt with by Mr Ian Clarke, 60 Oak Way, Feltham, Middlezex.

TR triumphant

An early production TR2 showing the air intake devoid of grille, the doors extending the full depth of the body and the wheel spats.

THE SUCCESSFUL RECORD

A thirty-year-old ancestor of the TR, the 1929 supercharged Triumph Super Seven Sports which competed in the T.T. that year.

TRIUMPH was a name of significance in the sporting world in the days before the Second World War, for as far back as 1929 Triumphs were competing in the Monte-Carlo Rally, in the Tourist Trophy race over the Ards Circuit in Northern Ireland and at Brooklands in single-seater form. A Triumph was the highest-placed British car in the 1930 Monte-Carlo Rally, won the Coupe des Dames driven by Mrs. Vaughan in the 1932 event and the 1,500 c.c. class in 1934 and 1935. Triumphs also competed with considerable success in the pre-war Alpine Trials and a Triumph saloon finished third in the general category and won the 1,500 c.c. class in the 1935 Liege-Rome-Liege Rally. It is perhaps somewhat ironic that many of the Triumph successes were the result of the driving skill or the organizing ability of a Cornishman named Donald Healey.

In 1939, however, the Triumph concern ran into financial difficulty—some hold through making too good a car for the price at which it was sold—and for a time the company was owned by Thos. W. Ward, the well-known Sheffield steel concern. In 1945, the company was bought by the Standard Motor Co., Ltd. and the first post-war cars to bear the Triumph name were the sleek 1800 Roadster and the razor-edge 1800 saloon later known as the Renown. Both were excellent cars, in fact many present-day owners are searching in vain for a roadster type of car with a hood which can be raised from the driving seat, or for a medium-size saloon of quality which is not just a quantity-production saloon with polished woodwork and a different radiator. Neither, however, was designed to appeal to the sporting driver, nor was the equally lamented Mayflower that followed, which was another case of a model that was only moderately successful when in production, but whose demise was deeply regretted.

In view, therefore, of the type of car built since the war by the Triumph company, most people were considerably surprised when a sports car made a last-minute appearance on the Triumph stand at the 1952 Motor Show at Earls Court. It can now be revealed that this new project was the result of a conversation between Christopher Jennings, the Editor of *The Motor*, and the then managing director of the Standard Company, Sir John Black. The Editor had recently returned from a visit to the United States, where he had been deeply impressed by the demand expressed on all sides for an inexpensive British sports car with

an engine of about two litres, for at that time a considerable gap existed between the 1,250 c.c. M.G. and the 3½-litre Jaguar XK120.

The Standard concern was already manufacturing the Vanguard engine of just over two litres in considerable quantity, but Sir John was very conscious of the fact that the Vanguard saloon was not the car with which to enter the American market. The idea of a sports car created round the Vanguard engine, however, seemed to be just what he was looking for, and he therefore listened with an eager ear to the Editor's remarks on the demand in America for just such a car.

A sports car fitted with the Vanguard engine was already being built in small quantities in this country and, as a first step, the Standard Company tried to buy the concern with a view to shifting production of the car to Canley or Banner Lane and there greatly increasing the output of it. However, the company concerned wanted to remain on its own and to continue building its individualistic cars in its own way, and many people will be

A grille for the air intake and external door handles mark the TR3, here shown fitted with the detachable hard top available as an optional extra.

The full width front grille and more sloping nose of the TR3A.

OF ONE OF BRITAIN'S BEST-SELLING SPORTS CARS

glad that this decision was taken, for the British motor industry has lost too many of its small but highly individualistic car companies since the war.

The Standard Company therefore set to work to produce its own sports car in the six months or so that remained before the opening of the 1952 Motor Show. The chief engineer regretfully abandoned any idea of a summer holiday that year and some very intensive work began. The task was made more difficult by the realization that the tooling cost of the new car must be kept as low as possible if it was to sell for a reasonable figure. This meant that as many existing components as possible would have to be employed, and that the body must also be relatively simple and incorporate no complicated curves which would push up the price of the body dies.

output soon showed that the target speed of 90 m.p.h. could not be obtained with a square-cut body and one of smooth aerodynamic form was therefore developed.

Drawings of the car were begun in late July, 1952, and construction of an actual car began only six weeks before the opening day of the Show. Nonetheless, by dint of tremendous non-stop work prototype cars were completed in time for display on the Triumph stand where they caused a great deal of interest. At the basic price of £555, it was obviously good value for money, but the looks of this ultra compact little car did not fill the average sports car enthusiast with enthusiasm, especially as the prototype had a sharply sloping tail occupied chiefly by the fuel tank whose filler formed the central carrier for the externally-mounted spare wheel.

Left: Distinguishing feature of the original Triumph TR1 which appeared at the 1952 Motor Show was its rounded tail carrying the externally-mounted spare wheel.

Right: This rear view of a TR3 shows the more shapely—and much more useful—tail fitted to all TR production models.

The target set the engineering staff was the production of a 90 m.p.h. car selling at a basic price of £500. It was expected that the car would be produced in relatively small numbers, which was another reason why the cost of tooling for it had to be kept as low as possible. The components finally selected from among those either already in production or which had been in production so recently that the tools and jigs still existed were a Vanguard engine with the bore reduced from 85 mm. to 83 mm. in order to bring down the capacity from 2,088 c.c. to 1,991 c.c., a Vanguard gearbox shorn of its steering column gear change linkage and equipped with a centrally mounted remote control, and with the number of speeds increased from three to four, Triumph Mayflower front suspension and rear axle and finally a chassis frame based on that of the pre-war Standard Eight of which production in its post-war form had finished in 1948.

At first it had been laid down that the body should be of the classic square-cut English type with separate wings and a slab rear petrol tank. A few calculations based on the expected engine

Moreover, people were all too apt to ask in that demanding tone that requires no answer: "What do Standards know about building sports cars?" The answer was "Practically nothing," but the company had sufficient intelligence to realize this fact, and to call to their aid in the development of the new model a man who had been working on fast cars for many years past, ever since the E.R.A. days, in fact. Ken Richardson joined the company in November, and at once took one of the prototypes to Lindley. He returned considerably depressed from this initial sojourn at what was to become almost his second home during the months ahead, for although the car showed promise in the matter of sheer performance, its steering, road-holding and braking left a great deal to be desired. At this point, the Standard Company might well have thanked Richardson politely for his expert opinion, but pointed out that questions of production made it impossible to carry out any fundamental alterations to the design. And if they had done so, the TR would have died very soon after its birth.

Ken Richardson and the TR2 in which, in May, 1953, he was timed at 124.889 m.p.h. at Jabbeke.

TR triumphant

J. Wallwork driving the winning TR2 in one of the special tests in the 1954 R.A.C. Rally.

Very fortunately, the company took no such attitude, but at once initiated a thorough-going revision of the design. A new frame of much greater torsional stiffness was produced, the front brakes were increased in size, the front wishbone links were modified and a wide variety of rear spring stiffnesses and damper settings were tried and rejected before a satisfactory combination was found.

The engine, too, underwent considerable development so that it eventually departed much further from the specification of the normal Vanguard engine than had at first been contemplated, but the result was an engine which is renowned for its trouble-free running even when thrashed for day after day in long-distance international rallies.

While all this work was proceeding on the mechanical specification of the car, the body was also completely re-designed at the rear, and the new lengthened tail not only improved the appearance of the car considerably but also provided more luggage room than even today is available in most sports car boots.

So swiftly had the work of development been carried out that the Triumph was able to appear in its revised TR2 form at the Geneva Show in March, 1953. In May of that year a cleaned-up version fitted with an under-shield, a metal cockpit cover and a small competition windscreen staggered most people by attaining 124.095 m.p.h. in a timed run on the Jabbeke motor road in Belgium. In a way, the car went too fast, for people were inclined to think that such an exalted speed from an inexpensive sports car of but two litres had been obtained only by super tuning, and that it therefore bore little relation to what the car would do as sold to the public. When the same car was run with the normal windscreen in place and the hood erect, its mean speed over the kilometre was 114.890 m.p.h., which is not so far off road test figures, and the difference could well be accounted for by the fact that the undershield was still in place and the car was minus its bumpers.

No further competition work was undertaken with the TR2 in 1953, largely because production did not get under way until July and then most of the early production models were dispatched to America. In 1954, however, it was a very different story. It is probably true to say that at the beginning of 1954 the Triumph TR2 was still regarded with a certain amount of suspicion by the keen rally drivers in Britain and in fact by the average keen clubman. A single event did much to change this attitude, the R.A.C. Rally. The 1954 Rally was one of the best of the whole series, tough but sporting, and out of 229 starters no fewer than 60 retired and only 7 retained clean sheets. The new TR2 in this its first major event finished first, second and fifth, the winner being driven by J. C. Wallwork. In addition, Triumphs won the Ladies' Prize, the 1,601-2,600 c.c. sports cars class and were runners-up for the team prize.

The following April the TR was given its introduction to international racing, for the works entered a car for the Mille Miglia to be driven by Maurice Gatsonides and Ken Richardson. It finished 28th out of some 450 starters, and a TR entered by Leslie Brook and driven by him and Jack Fairman also finished in spite of running out of petrol in the closing stages of the race. A third, privately entered, TR crashed.

It was in June, 1954 that the TR made its first appearance at Le Mans. The car was privately owned and was driven by E. B. Wadsworth and J. H. B. Dickson, and it looked so much the standard sports car among all the fierce racers that its appearance was greeted with a certain tolerant amusement by the vast crowd. However, to the surprise of everyone, it still kept going when a high proportion of the entry was no longer mobile, and it finally finished 15th in the general classification and fifth in its class at an average speed of 74.71 m.p.h., and with the staggeringly good average fuel consumption of 34.688 m.p.g. for the entire race. The regular high-speed running of this very inexpensive sports car and its amateur crew greatly impressed all who saw it,

Gatsonides and Bourelly finishing the 1955 Liege-Rome-Liege Rally, a tough Continental event in which TRs have been placed consistently.

Dickson at the wheel of the TR2 which ran so consistently in the 1954 Le Mans 24-hour race. Beside him is his partner Wadsworth, and behind him, Tommy Cain who ran the pit.

and ever since the TR has been most popular with Continental drivers.

In July that year the TR made its debut in the Alpine Rally in sensational fashion, winning the team prize and finishing second, third and fourth in the 1,600-2,000 c.c. class. Moreover, Gatsonides and Slotemaker gained a coveted Alpine Cup with their TR.

To round off an astonishing first season, two teams of three cars were entered for the T.T. on the tough Dundrod circuit, and all six cars finished, one team gaining the team prize and the other team being the runners-up for this award. No wonder Triumph celebrated by holding a victory banquet at the end of the year.

Since that astonishing first season the TR has won many other notable victories. In 1955 it won the British Mobilgas Economy Run driven by Dick Bensted-Smith of *The Motor*, it finished first, second and third in its class in the Liege-Rome-Liege Rally, and it finished first, second, third, fifth and sixth in the Circuit of Ireland Rally. In the 1956 Alpine Rally Triumphs won the team prize for the second time in succession (there was no Alpine in 1955), and five TR drivers gained Alpine Cups. Triumphs also won the team prize in the 1957 Liege-Rome-Liege Rally and finished third, fifth and ninth in the general classification. Miss Annie Soisbault, who joined the Triumph works team in that year, won the Coupe de Dames in both the Rally of Corsica and the Tour de France, and in 1958 won the Coupe des Dames in the Lyons-Charbonniere, Acropolis and International German Rallies.

Quite apart from its many successes in international events, of which the above is but the briefest selection, the TR has become almost the standard club sports car in this country. Just as impressive is the number of European rally drivers who compete with Triumph TRs. However, the car was designed and produced originally for sale in America; has it fulfilled this function? By February, 1959, 45,100 Triumph TRs had been built, of which no fewer than 30,000 had been shipped to the United States, which seems to answer that question pretty satisfactorily. The annual invasion of Britain and the Continent by plane loads of keen American TR owners to take delivery of their new cars is now one of the recognized features of the motoring year. These welcome invasions are organized by the American branches of the Triumph Sports Owners' Association, which now has some 6,000 members scattered throughout the world.

The development of the car itself has also been carried on with good effect since its introduction in TR2 form. In October, 1954, shallower doors were fitted with a fixed sill below them in place of the doors extending the full depth of the body which had proved difficult to open when the car was parked close to a kerb. In October, 1955, the TR2 was replaced by the TR3, which was distinguished from its predecessor by the addition of a grille to the previous unadorned air intake. A scuttle ventilator was also now incorporated and bigger S.U. carburetters increased the maximum power output from 90 to 95 b.h.p. Slight modifications to the cylinder head early in 1956 increased this figure yet again to 100 b.h.p. When a team of three TRs were run by the works at Le Mans in 1955—and all three finished—two were fitted with disc brakes for the front wheels. In October, 1956, the TR became the first sports car to be available with disc brakes for the front wheels as standard equipment and without any increase in price. The latest development was the introduction in January, 1958, of the TR3A with its full-width radiator grille, reshaped nose and a number of other improvements.

And there, for the present, the matter rests. The company has announced that a twin overhead camshaft engine is under development, but even without such an addition to its muscles the Triumph TR is still, at this interval of time since its introduction, one of the cheapest 100 m.p.h. cars made in Britain

The World Copyright of this article is strictly reserved
© *Temple Press Limited, 1959*

Triumphs have done consistently well in Alpine rallies, ever since the pre-war days. A TR is here seen in action on the Col Mojstrovka in Yugoslavia.

PROFILE

The TR2's strength soon made it a popular rally machine – this is Ken Richardson climbing a pass on the 1954 Alpine Rally

JOYFUL AND TRIUMPHANT

The early Triumph TRs – 2, 3 and 3A – are the epitome of rugged and entertaining British sports car motoring in the fifties. Graham Robson and Giles Chapman are your guides

Here is the shortened version of a classic fairytale – that of the ugly duckling which turned into a swan. In the beginning, there was a stubby little sports car prototype of 1952, without a pedigree, from a company which knew nothing about building sports cars, and in the end there were the best-selling Triumph TR2s, TR3s and TR3As, which in one bound put Standard-Triumph into direct and successful competition with MG and Austin-Healey. Let no-one forget that this was the car, no more and no less, which eventually persuaded the board of Standard-Triumph that their future cars should *all* be badged 'Triumph', and that the long-established 'Standard' marque name should be dropped completely.

It's easy to look back at these TRs, with their rugged, no-nonsense engineering and their proud sales *and* competition record, and forget that they were conjured up in a matter of months, put into production by a company whose previous 'sports car' had been the bulbous and old-fashioned 1800 Roadster, and sold in the USA through a small and inexperienced dealer network. Yet *this* was one of the three British models (the MGA and the Austin-Healey 100 were the others) which completely changed the image and reputation of British sports cars in the USA, and helped to bring about a sales boom which lasted well into the sixties and seventies. A mixture of bureaucracy and BL incompetence ended it all.

Perhaps you still can't understand what all the fuss was about, not unless you were actually a red-blooded young motorist of the fifties. Just consider: in 1952, if you wanted to buy a British sports car, you first had to wait patiently in a queue (there were waiting lists for *everything* then), you had to raise the money, and finally you had to face up to a rather restricted choice. There were the MG TD, the Morgan Plus Four, and the Jowett Jupiter – all worthy and interesting, but none of them very fast – and there was the Jaguar XK120, which was a great performer but also very expensive. Between these two extremes there was a large gap.

Then, in that significant year, MG designed the prototype of the MGA (although it didn't appear in public until 1955), Donald Healey evolved the Healey 100 (which became the Austin-Healey 100 of 1953), and Standard-Triumph rushed through the design of their prototype '20TS' sports car.

The 20TS, let's face it, would never have made any money for Standard-Triumph if they had put it on sale. It had a flexible chassis (the first project schemes even considered using surplus 1936-style Standard Flying Nine frames!), an under-powered 75bhp engine, and strange styling which combined a modern nose and centre section with a 'traditional' tail incorporating an exposed spare wheel, a fuel filler cap poking out through the centre of that wheel mounting, and no enclosed luggage accommodation.

Fortunately, Standard-Triumph reacted to distinctly cool media and public response at the Earls Court Motor Show, hired Ken Richardson as their chief sports car development engineer, and a team of engineers led by Harry Webster produced a stiffer new chassis, an engine up-gunned to 90bhp with amazing longevity *and* fuel economy, and a revised rear bodystyle with an enclosed spare wheel and a separate luggage boot. The much-improved car, now called the TR2, sold for £787 at a time when the MG TD cost £752, the Morgan Plus Four £802, the Austin-Healey 100 £1064, the Jupiter £1127 and the XK120 £1602. It was a very saleable proposition.

Standard-Triumph did a great job in getting the revised car on sale a mere five months after the first public showing, and although the new car was remarkable in some respects, there was also a lot wrong with it. For its price, and considering its specification, it was a very fast little car (at 103mph, it was 20mph faster than the MG TF), and there was no doubt about its economy, for 32-35mpg was normal for many owners and Richard Bensted-Smith used his own car to record no less than 71mpg in the no-holds-barred Mobilgas Economy Run of 1954.

But it was very noisy at first, with that characteristic exhaust 'bark' in mid-range. It had rather dodgy, front-biased drum brakes, and handling quirks which took a lot of learning. The handling problem was twofold: there was the basic design problem in which the back axle rode *above* the chassis side members and soon ran out of movement in cornering conditions, and, secondly, this made very hard spring and damper settings desirable. The result was a car which skipped merrily from bump to bump in a bouncy, crashy way, and on the rudimentary tyres of the period it was all too easy to induce breakaway at the front *or* rear. Michelin X tyres improved overall grip just as long as the treads were in contact with the road, but the early Xs had vicious on-the-limit breakaway, especially in the wet …

TR2s went racing: at Le Mans '55 (from left) are Dickson/Sanderson, Goodall/Broome, Haddeley/Richardson

Most TR owners raced or rallied their cars, and took home masses of trophies as long as they kept on the straight and narrow

Not that this harmed the TR's early reputation at all, for it offered such phenomenal value for money that its long-suffering owners forgave it everything. It soon proved to be such a rugged car that it began to dominate the British and European rally scene. In a TR you could rush over all manner of road surfaces and not damage the underside of the chassis. In an MG TF there was simply not enough performance to meet demanding schedules, and in the Austin-Healey the exhaust system and the rear of the bodyshell was always at risk. Most TR owners rallied or raced their cars, and took home masses of trophies as long as they kept the stubby little machines on the straight and narrow.

There was criticism from some quarters, not least from MG enthusiasts who complained that the TR2s and TR3s had no 'pedigree' (whatever that meant), and from Austin-Healey fans who suggested that the TR's styling did not compare with the smooth lines of the 100/4 (or the later six-cylinder-engined Big Healeys), but this did not seem to deter the customers. Triumph went through a dodgy patch in 1954 and '55, when it took time to establish the TR's reputation in the USA, but sales built up rapidly thereafter, and the TR's name and future were assured.

The early TR2 needed improvement to make it more civilised and versatile, but you could say that this was all part of the learning process which the design and sales team carried out very speedily indeed. In every respect except fuel economy, which worsened gradually but noticeably with each new model, the TR seemed to improve yearly, although the performance stuck at about the same level for eight years.

There wasn't much that anyone could do about the roadholding – changes to the chassis design were ruled out as an 'over-axle' side member design would have been too costly to tool up – but the engine's torquey reliability allied to the optional overdrive transmission made up for that. When front disc brakes were adopted from the autumn of 1956, a good car became a great one. Almost every British rally, it seemed, was won by a TR, and on some national events there would be 20 or 30 cars in the entry lists.

Even in its original form, the TR2 was well-equipped, for the hood and side screens were weatherproof in all conditions, the bucket seats were comfortable and there was ample leg room. For a company without sports car 'pedigree', Standard-Triumph had produced a clear and informative facia panel display, a splendid gear change, and a very effective fly-off handbrake, all allied to more luggage accommodation than the competition.

TR2s went rallying: on the Alpine in '54 (from left) are Richardson, Slotemaker, Gatsonides and Kat

TR3, produced 1955-57, is distinguished by new grille   *Practical rear seat became an option with the TR3*

SPECIFICATION Triumph TR2

Engine	In-line four
Bore/stroke	83mm × 92mm
Capacity	1991cc
Valves	Pushrod ohv
Compression	8.5:1
Power	90bhp (gross) at 4800rpm
Torque	117lbs ft at 3000rpm
Transmission	Four speed manual, optional overdrive on top, third and second gears (early cars on top gear only)
Final drive	3.7:1
Suspension front	Ind by wishbones, coil springs, telescopic dampers
Suspension rear	Live axle, by half-elliptic leaf springs, lever-arm dampers
Steering	Cam and lever
Brakes	Drum/drum, no servo

DIMENSIONS

Length	12ft 7ins
Width	4ft 7.5ins
Height	4ft 2ins
Weight	1848lbs (with no extras fitted)

PERFORMANCE

Max speed	103mph
0-60mph	11.9s
Standing ¼-mile	18.7s
Typical mpg	33

Once the model improvements and the list of optional extras were made available, the later TRs were even better bargains. There was, of course, a body rust problem as the cars got older, but many of those rusty panels were bolted-on, rather than welded-in, items, and could be easily renewed. All the best TRs had centre-lock wire wheels, Laycock overdrives (the switch being on the panel, close to the steering wheel rim) and a removable hardtop, while a few even had the 2.2-litre engine which Ken Richardson's 'works' rally team first used in the Alpine Rally of 1958.

We must not forget that the 1957 model TR3 was the first British series production car to have disc brakes (the Jensen 541R and sports racing machines like the Lotus 11 really don't count as 'series' production models), and we must also remember that the TR3A outsold the competition from Abingdon in the commercially vital USA market in the late fifties.

The tragedy today is that so few of these cars still exist in top quality restored condition in this country. Many of the older TRs have been 'converted' to TR3A styling by the use of the later-type front panel and grille, and a plain-grille TR2 is thus a rarity, except at TR Register meetings. It's easy, therefore, to forget what a *small* car the TR2 was (at 12ft 7ins long, it is 7ins shorter than a Ford Escort), and it combined that chunky styling with such useful two seater accommodation.

It was also the original – they don't make them like that any more. Somehow, all the TRs which followed were an anti-climax.

Production History

Although the original Triumph 2-litre sports prototype, coded 20TS by the factory, and never *officially* known as TR1, was unveiled in October 1952, the first 'off-tools' production cars were not completed until the summer of 1953. Indeed, the 20TS ('TR1') design, complete with its flexible chassis, short tail and exposed spare wheel, was not used in production, the revised model being shown at the Geneva Show of March 1953. Austin-Healey fanatics, incidentally, have always liked to boast that the 100/4 was revealed at the same Earls Court Show in prototype form, but that deliveries began in the spring of 1953, well before the TR2 went on sale …

Lots of dials – cockpit was little changed in nine years

The re-developed Triumph sports car was given the definitive title of TR2, and went on to the market, rather haltingly, in the summer/autumn of 1953. In that calendar year, incidentally, only 248 cars were completed, of which just 50 stayed in the UK. Volume sales, almost all to export territories, began early in 1954.

The first cars were delivered in 'basic' trim, which is to say that no optional extras were then available, but from early 1954 several important options were added to the catalogue. In particular, these included Laycock overdrive, centre-lock wire wheels, and Michelin X radial-ply tyres, the overdrive originally working only on top gear.

Then, in the autumn of 1954, several significant improvements were phased in. The body structure was changed to what is commonly known as the 'short door' configuration (to allow the doors to be opened over high kerbs!), the rear brakes were increased in size and effectiveness, and overdrive

BUYERS' SPOT CHECK

Buying a TR2/3/3A as your classic choice makes sense: these cars are tough, dependable, relatively inexpensive to buy and rebuild, cheap to maintain, fast enough not to be an embarrassment in modern traffic and, provided you do not mind a spot of leakage in rain and a few draughts, perfectly practical to use every day. The family man can even get two small children into the space behind the seats in tolerable comfort. Yet despite its usefulness, the car is a classic sports roadster in the great tradition, all wind in the hair and cutaway doors!

In the late sixties and seventies, the 'sidescreen' TRs were somewhat undervalued, and did not always get the cars and maintenance they deserved, many examples being driven into the ground by the impecunious looking for a fast, eye-catching car at minimal expense. The situation has changed today, for the better, and enthusiasts are now appreciating the cars for their excellent qualities, and spending money to preserve and renovate them. This in turn has led to the establishment of a large number of TR specialists, offering both spares supply and restoration, and although there are exceptions, these specialists are mainly helpful and knowledgeable. Their existence, coupled with support from the TR Register, has enabled many enthusiasts to undertake complete rebuilds that would otherwise have proved impossible, at a realistic cost.

The TR buyer has effectively three choices. First, the non-running car in poor condition, of which there is still a reasonable supply; secondly, a roadworthy car that has probably had some recent work to keep it operational, these being available at around five times the cost of the non-runner. Such a vehicle will, however, need further expenditure, especially to bring it to perfectionist standards. The third choice is to buy a fully restored car, or if one can be found, an original and fully maintained example.

Clearly, the latter course is only open to the reasonably affluent, though a TR fancier who either lacks time or mechanical knowledge would probably be better off with such a vehicle, rather than one from the first two categories, and would most likely find it cheaper in the long run. The classified columns in the TR club magazines are a good starting place to find cars for sale, for the vendor will frequently be an enthusiast, and prices tend to be more realistic than elsewhere. There are specialist TR dealers, although inevitably they tend to be more expensive, and concentrate on later TRs as the supply of vehicles is better.

In terms of parts costs, insurance and fuel economy, the TR compares favourably with modern cars, with the added advantage of no depreciation. The car's main virtue is its strength, but even the youngest 3As are 25 years old, so inspect thoroughly.

For the amateur restorer, the separate chassis frame is a great asset compared with a monocoque design, and the chassis rarely suffers terminal rusting, though the cross-tubes can rot. Replacements are available, as are new body support outriggers, arguably the weakest point on the chassis. The box-section main members are usually sound, but sometimes rust where they pass under the axle and turn up slightly towards the rear. The front of the chassis is usually well-preserved by oil-leakage, although the small cross tube at the front sometimes rots. The chassis should be checked for accident damage, particularly at the front, for misalignment around the front suspension pillars can cause tracking and handling difficulties.

The front suspension is conventional, with double wishbones connected by a vertical link carrying the stub axle. The link swivels on a top ball-joint and a socket and trunnion arrangement at the base. Coil springs are used, and the whole suspension can be easily overhauled without special tools, all moving parts being obtainable except the vertical link itself. This causes problems in that the thread that fits into the trunnion socket can wear, and replacement of the trunnion alone can be insufficient to eliminate play for MoT and safety purposes.

In terms of parts costs, insurance and fuel economy, the TR compares favourably with modern cars, with the advantage of no depreciation

Check the suspension carefully for wear, for although straightforward, it contains many individual parts and costs can mount up. Regular greasing is necessary with this design to obviate stiffness. Play in the steering can usually be adjusted away, and while new idler arms/brackets can be purchased, replacement steering boxes cannot. Parts to recondition are available, but it is as well to remember that TR steering was not wonderful when new, and does not compare in feel or lightness with a rack and pinion system.

The rear suspension is both basic and strong, the axle being located solely by two semi-elliptic springs. TR2/3s had the Lockheed-braked axle, which is prone to half-shaft breakage and hub seal leakage, but the later Girling-braked cars had a redesigned axle eradicating these faults. Many earlier examples have had the later axle fitted, identifiable by its having six backplate mounting bolts instead of four. Rear dampers are lever arm, easily and cheaply replaced, but the mounting brackets can fracture. The crown wheel and pinion, differential and propshaft are well up to their tasks, rarely giving trouble.

The engine is famous for its robustness, and is a wet-liner design based on a Ferguson tractor unit. Tappet noise is common, and not generally of concern. Bore wear is easily dealt with by replacement piston/liner sets, and many TRs have had their capacity uprated by the fitting of 86 or 87mm pistons, giving 2138 or 2187cc. This modification gives much improved torque, and has no disadvantages. Oil pressure should be 60/70psi hot at 2500rpm, but on idling can drop to 15/20psi. Occasional crankshaft breakage occurs, probably the only major problem to afflict the TR motor. It is very rare, however, and the car can usually be driven home, so tough is this engine! Bearing life is good, and engines can run for 100,000 miles without dismantling, given regular oil changes.

The gearbox is equally strong, likewise the optional overdrive unit. Malfunction here is usually confined to dirty or insufficient oil. or relay or solenoid problems, both easily rectified. Clutch problems are rare, although it can stick on its splines on a car that has stood for some time. This can usually be cured externally by the application of force! Brakes are straightforward, though the early front drum system can fade under hard usage. The disc brakes are excellent, and have no servo to go wrong. Virtually all brake parts are available, the exception being the combined master brake and clutch cylinder on drum-braked cars. These are extinct now, but reconditioning kits can be purchased.

Bodywork condition is paramount in assessing the value of a TR, for rust can attack anywhere, and very few TRs/3/3A survive that have not had at least one body rebuild. Fortunately, most of the panels are available new today, although these replacement panels can be trickier to fit than the unobtainable Stanpart originals. Floors, inner and outer sills, wings, door bottoms, spare wheel pans and boot floors all rust but can be purchased; boot lids and front apron panels can be more difficult, though some remanufacture has occurred in this area. 'A' posts rot through at the base, as do the quarter panels behind the 'B' posts and the rear inner wings. Panel prices are reasonable compared with other classics, and the mechanics of body restoration are simpler than on some contemporaries, Austin-Healeys for instance. Any purported fully-rebuilt car should be thoroughly checked to see that it is indeed so, correct door gaps being a good indicator to the accuracy of the work.

New weather equipment is readily available, although only the later 'Dzus' type sliding sidescreens have been remanufactured. Seats are difficult, and it is essential to have the frames at least so that specialist renovation can take place. If the original seats are missing, finding replacements in any condition can be hard. Other items of trim can usually be found secondhand or obtained from the trimmers who specialise in the TR. As regards interior fitments, switches and instruments are reliable and if missing can be found secondhand easily, but the original steering wheel and horn push boss are very difficult to come by.

Bill Piggott

was re-arranged to operate on top, third *and* second gears. Not only that, but a smart removable hardtop also became optional. This, therefore, became the definitive TR2, which was fast and economical, but still possessed rather unruly roadholding and an exuberant exhaust note.

The TR2 gave way to the TR3 from the autumn of 1955, a point at which the only important changes were to the engine (95bhp instead of 90bhp) and the nose, where an 'egg-box' grille was fitted. At the same time an optional '+2' rear seat was made available, though this was of little practical use, as the leg room was virtually non-existent. During 1956 there was a confusing period when not one but three different cylinder head castings were fitted to production cars, but a power boost to 100bhp was not finally achieved on *all* production cars until the summer of 1956.

The next important development came in the autumn of 1956, when the original Lockheed brakes were discarded in favour of a new Girling system with front wheel discs – the very first such installation on a British production car. At the same time, a significantly stronger back axle (Vanguard III instead of Mayflower-type) was standardised. There was also a rather rudimentary 'GT' kit as an option (to allow the TR3 to be used in 'closed GT' classes in rallies), which added a steel hardtop and exterior door handles to the basic specification.

The TR3 had become the TR3A for 1958, complete with full-width, 'dollar grin' front grille

This was really the end of important mechanical changes to the specification of the classic TR, for in the next few years all the headlines were made by ever-increasing sales. In 1955, just 4399 TRs had been delivered, but in 1957 this rose to 10,598 cars, in 1958 to 15,996, and in 1959 (the peak year) to 21,298. After that, there was a steep decline (for unsold stocks in the USA were growing fast), and assembly died out in the spring of 1961.

Along the way, though, the TR3 had become the TR3A for 1958, complete with full-width 'dollar-grin' front grille, a lockable boot lid and exterior door handles as standard, while from 1959 an optional 2.2-litre engine was offered (large-bore liners and pistons were really the only changes). In the same year, the body tooling was substantially renewed, with significant panel changes under the skin.

In theory, Triumph management thought that the old TR should then have been replaced for 1962 by the new Michelotti-styled TR4, but the USA distributors wanted to hedge their bets and asked for the old-shape car to be continued as well! This was difficult to arrange, but in the end a revised car, now known as TR3B (though never badged as such), was assembled at the Triumph-owned Forward Radiator Co factory in Birmingham during 1962 – all previous TRs, and the new TR4, were assembled at the Canley, Coventry complex.

All TR3Bs were exported to the USA, and they all had left-hand drive. There were two chassis number series – the first 500 cars, with TSF Commission Numbers, had the 1991cc engine, and those with TCF numbers had the 2138cc engines – but both types had the new all-synchromesh TR4-type gearbox which was never found on any of the earlier TRs. A few of the early TSF cars may originally have had non-TR4 gearboxes. Confused? So are the records!

Production Figures

TR2	1953-55	8628
TR3	1955-57	13,377
TR3A	1957-61	58,236
TR3B (USA sale only)	1962	3331
Total	**1953-62**	**83,572**

Home market sales: 2823 TR2s, 1286 TR3s and 1896 TR3As; a total of 6005 cars in eight years

Clubs, Specialists & Books

The age threshold for a car to be considered classic is advancing all the time, with many cars from the late seventies and even early eighties now having an enthusiastic following. With cars like the Lotus Esprit, for example, membership of an owners club may not be particularly necessary, but if you own, or are thinking of owning, a Triumph TR2, 3 or 3A, then getting together with other owners and fans of these sometimes 30-year-old cars is very wise.

There are two clubs open to you. The largest is the TR Register, which caters for all of the sporting Triumphs up to the TR8. Formed in 1970 and with almost 5000 members, it costs £14.00 per year to join, for which you will receive eight copies of *TRaction* annually, together with access to beneficial discount and agreed value insurance schemes. Being RAC-affiliated, the TR Register takes part in many events around the country, as well as running their own race championship, and joining up with TR clubs abroad. For details, contact Valerie Simpson at the TR Register Limited, 271 High Street, Berkhamsted, Hertfordshire (tel: 04427 5906).

The other club is the TR Drivers' Club, formed in 1982, which now has a membership comfortably exceeding four figures. They have their own discount arrangements with selected suppliers as well as having a deal giving favourable insurance rates. The TR Drivers' Club race series takes place each year, but there are also many other social events in their calendar, so lovers of noggins, natters, driving tests, autojumbles and concours will not be disappointed. The club magazine, *The TR Driver*, is published every other month, and it's included in the annual £10.00 (for UK) membership fee. For further information, contact Steve Clare, 22 St Martin's Drive, Walton on Thames, Surrey. Both TR clubs have regional sections spread all over the country, so there is sure to be a local chapter near you.

There are many specialists in Britain who will be able to help with everything from supplying the odd trim part to full restorations, and as a guide for ready-reference, here are some of the major ones.

Cox & Buckles, 22-28 Manor Road, Richmond, Surrey TW9 1YB (01 948 6666) and also 89 Fairfax Road, West Heath, Birmingham B31 (021 477 7966); Churchfields, 214 Hatfield Road, St Albans, Hertfordshire (0727 40181/2); Northern TR Centre, Sedgefield Industrial Estate, Sedgefield, Cleveland TS21 3EE (0740 21447); TR Shop Limited, 16 Chiswick High Road, London W4 (01 995 6621); TR Improvements, 19 Caernarvan Road, South Woodford, London E18 (01 505 3017); Dee and Gee Auto Repair Centre, 19 Lime Tree Road, Washwood Heath, Birmingham (021 327 5211); Motor Hoods (Colchester) Ltd, Unit 1, Grange Farm Road, Whitehall Industrial Estate, Colchester, Essex (0206 70902); John Skinner Retrimming Specialist, Unit 2, Carpenters Buildings, The Avenue, Cirencester, Gloucestershire (0285 67410); M.E. & J.W. Pumford, Unit 4, Corporation Road, Birkenhead, Merseyside (051 653 4313); TR Enterprise, Haywood Oaks, Blidworth, Nottinghamshire (0623 793 807); Triumph Care, Crown Works, 1 Church Road, Norbiton, Kingston upon Thames, Surrey KT1 3OB (01 549 9305); Nicol Transmissions, Coppice Trading Estate, Stourport Road, Kidderminster, Worcestershire DY11 7QY (0562 752651); Autogear Transmissions, 467 High Street, Leytonstone, London E11 4JU (0268 681680); TR Spares South West, The Garden Cottage, Grove Hill, near Bridgwater, Somerset TA7 0GJ.

If you are tempted by the traditional lines and raunchy performance of an early TR and would like to gen up on them further, then there is plenty more reading matter around to satisfy your inquisition. Chris Harvey's *TR for Triumph* is perhaps the weightiest tome on the subject, and it's well written and researched for the price of £14.95. Triumph expert Graham Robson has penned one of the excellent *Collector's Guide* books on all the TR Triumphs and it is very good value at £8.95. Contemporary road tests of these truly classic sports cars are reprinted in two Brooklands books, one on Triumph sports cars generally from 1953 to 1967, and one solely of articles that appeared in *Road & Track* from 1945. Finally, as if that was not enough, Bill Piggott, featured in our 'Owner's View', is preparing a Haynes *Super Profile* on the TR2, 3 and 3A. It will be out for Christmas.

Ken Richardson prepares TR3As for the Monte Carlo

GT kit for TR3 included top and sliding windows

TR3A had a full-width grille and lockable boot

TR2 was an arduous – if uncomfortable – rally car

RIVALS WHEN NEW

The TR2 and 3 were of that peculiarly British genre of motor cars which did not have rivals from abroad, and perhaps this is why they proved so popular in export markets such as the USA. Most Italian sports cars of the time were based on far more refined engineering principles than the Triumphs, principles that would never have allowed for the use of a tractor engine and a chassis based on that of a particularly dull family car of two decades earlier.

Sporting cars in Germany were few and far between of any sort; most people could not afford anything as expensive as a BMW, Mercedes or Veritas, and a Porsche, even in the early fifties, was well out of reach of most wallets. France's only sports cars were either air-cooled miniscules or dying and overbodied pre-war *grandes routiers*. With irrelevant and remote exceptions (the Woodill Wildfire …?), sports cars weren't made anywhere else in the world.

In Britain, however, the situation was different. Partly due to the demands and enforced pressures of the export industry and partly to the type of car the British had *always* made, there were a number of reasonably priced and reasonably fast small sports cars available at the time of the TR2's introduction in the closing months of 1953. The Austin Healey 100/4, with its big Austin four-cylinder engine, race-bred chassis and stunningly masculine good looks, was perhaps the most obvious competitor, although it was both more powerful and more expensive. The MG TF, despite its similarly updated pre-war style, was no match for the Triumph in the performance stakes – it was more at home outside the Olde English Thatched Cottage than thundering along surrounding country lanes.

A rather more upmarket alternative to the TR2 was the Sunbeam Alpine, the ST 90-based two-seater roadster that was made in limited numbers between 1953-55. Competition in slightly more vintage style could be had from either the HRG 1500 (still *just* in production) or the Vanguard-engined Morgan Plus Four (still very much in production). The Swallow Doretti was really nothing more than a coachbuilt TR2 anyway, with a more civilised body and interior, so this car was not so much a rival, more a complementary model. The Jowett Jupiter was also a fine sporting car, once one got over the idiosyncracies of its flat-four engine, and provided the TR2 with some stiff rally competition in its last years.

By the time the TR3 was announced, the number of rivals had diminished considerably, as the Alpine, TF, Jupiter and HRG had all gone. However, there was now the sophisticated 1.5-litre MGA, with its good looks and fine chassis (not forgetting the potent but troublesome twin cam MGA roadster) and if acceleration was an important factor, then one could have done worse than the Fairthorpe Zeta with its 2.6-litre Zephyr engine and lightweight body. This car, together with the smaller Coventry-Climax-engined Electron, were popular club racers in the late fifties but suffered the ignominy of bizarre and outstandingly crude bodies.

PRICES

Unlike some other classic cars, there is no premium to be paid for either an early or a late TR2, 3 or 3A, as the values for these cars are the same regardless of model. You might be surprised to learn, however, that an example of this enchanting roadster, albeit in off-the-road, dismantled condition and needing a total rebuild, can be picked up for as little as £300-400, a car requiring slightly less attention probably setting you back £800.

A good everyday car with an MoT, something usable but by no means perfect, will fetch from £2000 to £3500, while excellent cars needing virtually no work can make up to £5000. For concours cars, expect to pay as much as £7000.

OWNER'S VIEW

Bill Piggott poses proudly by his almost-running Triumph TR2, which he has built up from eight other cars!

Bill Piggott is all set to enjoy the simple pleasures of TR2 motoring this summer, when his car finally hits the road

The TR2 and TR3 are such rugged and reliable cars that it would be practical to use them as everyday transport, for both driving to work and the underground car park during the week, then sprinting up Prescott and whizzing along sunny country roads at the weekend. If, however, you are anything like Bill Piggott, with a wife, growing children and a successful, home-run business as a solicitor, then you will probably need a hefty Volvo estate car too, but driving this rather pedestrian vehicle when it's raining has not diminished his love of Triumph TR motoring one bit.

Bill Piggott's affair with the TR goes back to 1970, when he bought his first TR2 for £50, and he recalls that it cost more than that just to insure the car. He was so keen on the car that he has owned literally dozens of them since, having at one time or another possessed every type of TR2 and 3 that has been made. His current car is an early-1954 'long door' TR2, registered TVT 35, and is reaching the end of an extensive home restoration, in readiness for those balmy August days to be spent in the Buckinghamshire countryside surrounding his home near Thame.

Although now looking very trim in its new coat of British Racing Green paint, the car could hardly be described as original, for Bill admits, with more than a little inverted pride, that parts from as many as eight cars have been put together to produce one well-sorted example.

"For me personally," said Bill, "a good, driveable car is more important than a totally original one, but all the same, most of the car is made from original parts. It hasn't been on the road for 18 months, but when the restoration is finished in a few weeks' time, it will be used for daily summer motoring. April to October, but taxed all year round so I can go to club meetings in it. As an all-round second car, there's room for the kids and plenty of luggage space."

Bill's car was bought as a non-runner, with a reasonable body but 'shot' mechanicals, and he has amalgamated it with most of another car. He managed to find all the parts that were needed to complete the car, such as the front bumper. The car was resprayed in the family garage using borrowed equipment, and the paint finish is excellent for a home job. He has also made quite a few alterations to the specification, like replacing the old 2-litre Vanguard engine with a 2.2-litre unit from the later saloons (he's hanging onto the old one, though, just in case he gets tired of the extra power!), fitting wire wheels and the very necessary overdrive. Other modifications include the fitting of an anti-roll bar and front Konis, and a single-box stainless steel exhaust system that gives that essential TR2 'roar'. The car already has the rare telescopic steering column. A TR3A axle is fitted. The aero screens add a sporting touch to the car, although unless you like a faceful of dead flies, they are slightly less practical for non-racing activities …

"For me personally, a good driveable car is more important than a totally original one"

We asked Bill what the attraction to the TR2 was for him: "They have always been very good value for money, and the car is a very good compromise of semi-pre-war style and decent post-war performance. Home maintenance is easy, spares are cheap. One of the best things about the car, though, is that it can more than keep up with the average modern car, and the TR's reputation of a tail-happy rear end is dispelled by the fitting of modern radial tyres in place of the original cross plies. TR hardtops, however, are a waste of time, because they make the car very noisy – even in winter, the hood's a better bet. Overdrive is virtually essential, a must."

Bill is the registrar of the TR2/3 section of the TR Register, and has kept historical and car survival records since 1979 for the club. Naturally, he would advise all those who own or want to own a TR to join the club. He told us that the TR Register is a club for people who want to enjoy driving their cars rather than those who want to spend all day polishing them at concours events.

Enjoyment of his car is something that is foremost in Bill Piggott's love of the early TR, and he feels that anyone who wants a gutsy fifties sports car with period looks and a rugged temperament could do little better than owning one.